MACH'S PHILOSOPHY OF SCIENCE

Mach's Philosophy of Science

J. BRADLEY

THE ATHLONE PRESS
of the University of London : 1971

Published by
THE ATHLONE PRESS
UNIVERSITY OF LONDON
at 2 Gower Street London WC1

Distributed by Tiptree Book Services Ltd
Tiptree, Essex

U.S.A.
Oxford University Press Inc
New York

0 485 11124 1

PRINTED IN GREAT BRITAIN
AT THE PITMAN PRESS, BATH

To my teachers,
the late Victor Oscar Newton
and the late Alexander Wood

Preface

My interest in Mach arose out of a lecture on the lever given to freshmen at Cambridge in 1927 by the late Alexander Wood. He advised his students to read Mach, and I have been doing so ever since.

This book is confined to one aspect of Mach's work, his occasional comments on the philosophy of physical science. These comments amount to a fairly complete elementary introduction to the philosophy of science, although Mach failed to include any discussion of questions of probability. Mach takes a rich sensuous experience as the basis of science; to him theoretical systematisation is secondary. I believe that Mach's assessment of science is an important one. The beginner may find the book a useful introduction.

Mach wrote three admirable textbooks. In the course of this account of Mach's philosophy, I have found it necessary to set out the foundations of dynamics and thermodynamics in Mach's own style. These summaries could be useful to the young student of physics. If they direct the more mature reader to Mach's texts, they will have been worth while.

There is no Mach Principle in the writings of Mach. Indeed, the so-called principle was not formulated until some years after Mach's death in 1916. Here I have traced out the evolution of this affair in a simple factual manner. This may be of some interest to physicists and philosophers.

The most important part of the book, in my view, is the classification of metrical concepts according to their relative 'closeness to' and 'remoteness from' sense-perception. This is Mach's own idea, but I have extended it somewhat.

Professor H. Dingle has been a constant inspiration and guide since about 1950. It is a pleasure to acknowledge the debt I owe to him, and to other physicists and philosophers. In particular I should like to thank Dr G. Burniston Brown, Dr G. J. Whitrow, Mr C. E. Reed, Dr Axel Stern and Mr C. Hamill. Further I must thank Mr W. R. Fraser for his great help in checking my translations and suggesting amendments where necessary.

Also I am most grateful to Mrs B. Mandeville for typing the whole book, and to my sister, Miss M. Bradley, for drawing the diagrams.

J.B.

Contents

MACH'S MOST IMPORTANT BOOKS, TRANSLATIONS AND REFERENCE LETTERS

A. *The Analysis of Sensations*, fifth German edition, translated C. M. Williams and S. Waterlow, Chicago and London: 1914. First English translation, 1897.
Die Analyse der Empfindungen, fünfte vermehrte Auflage, Jena: 1906. First German edition, 1886.

E. *Erkenntnis und Irrtum*, zweite Auflage, Leipzig: 1906. First edition, 1905.

G. *The History and the Root of the Principle of the Conservation of Energy*, translated P. E. B. Jourdain, Chicago: 1911.

GG. *Die Geschichte und die Wurzel des Satzes von der Erhaltung der Arbeit*, zweite Auflage, Leipzig: 1909. First German edition, 1872.

M. *The Science of Mechanics*, ninth German edition, translated T. J. McCormack, Illinois: 1960. First English translation, 1893.
Die Mechanik, neunte Auflage, Darmstadt: 1963. First German edition, 1883.

O. *The Principles of Physical Optics*, translated J. S. Anderson and A. F. A. Young, London: 1926.
Die Prinzipien der physikalischen Optik, Leipzig: 1921.

P. *Popular Scientific Lectures*, fifth edition, translated T. J. McCormack, Illinois: 1943. Earlier translations, 1894–98. This is a translation of *a part of* PP.

PP. *Populär-Wissenchaftliche Vorlesungen*, vierte Auflage, Leipzig: 1910. This collection first published complete, 1896.

S. *Space and Geometry*, translated T. J. McCormack, Illinois: 1943. This is a translation of pp. 337–422 of E. First English edition, 1906.

W. *Die Prinzipien der Wärmelehre*, erste Auflage, Leipzig: 1896. A second edition of W. (1900) was consulted but not quoted.

Official translations were used generally; other translations from German revised and checked by Mr W. R. Fraser.

Errata

p. 41 footnote: *for* O., p. 17. *read* [1]O., p. 17.

p. 49 line 15: *for* that *read* than

p. 58 line 7: *for* prfeerred *read* preferred

p. 69 line 32: *for of read* of

p. 72 line 18: *for* ohe *read* one

p. 76 footnote: *for* iteral *read* literal

p. 98 last line: *read* judged

p. 199 footnote: *for* [2]W., pp. 358–9. *read* [3]W., pp. 358–9.

p. 206 lines 5–6: *for* metalphysical *read* meta-physical

1 Sense-perceptions

1. Descartes and Mach

Descartes' first principle is man's duty to doubt everything that can be doubted, before the principles of philosophy are established. It is necessary for a man to pass beyond this condition of doubt, but it is also necessary for him to pass through it.[1] This at any rate, thought Mach, is an admirable start:

> The maxim of doubting everything that has hitherto passed for established truth cannot be rated too high; although it was more observed and exploited by his followers than by himself.[2]

After a remarkable analytical discussion, the details of which need not concern us here, Descartes is left with *two orders of substance*,

extended substance

and

thinking substance,

which he regards as distinct and separate, and concerning the being of which he can entertain no doubt. Mach, who refers to himself as a *monist*,[3] rejects this Cartesian dualism, although it is certainly not evidently true that *body* or *extended substance* is identical with *mind* or *thinking substance*.

Following Galileo,[4] Descartes considers that extended substance bears two main kinds of quality, the so-called primary and secondary qualities. The primary qualities are objective, *they are out there in nature*. In Descartes' form of the doctrine of qualities, the primary ones are exclusively mathematical and

[1] R. Descartes, *Descartes Selections*, ed. R. M. Eaton, New York: 1927, p. 12.
[2] M., p. 362. [3] A., p. 14.
[4] A. N. Whitehead, *Essays in Science and Philosophy*, London: 1948, pp. 173–4.

kinematic, viz. extension in three dimensions, figure or geometrical pattern, motion and number.[5] Secondary qualities are generated in us by the primary qualities actually there, acting upon and through our sense organs. It is an interesting fact that Descartes regards not only smells, colours, feelings of hardness and softness, and tastes as secondary; he also puts weight in the same category. Mach does not notice this. Neither Descartes nor Leibniz distinguished mass from weight. Locke and Newton usefully extended the doctrine of qualities by taking mass as an additional primary quality.

Mach has a just appreciation of Descartes' ideas on the primary mathematical qualities:

For Descartes, who in his opposition to the occult qualities of Scholasticism would grant no other properties to matter than *extension* and *motion*, sought to reduce all mechanics and physics to a geometry of motions, on the assumption of a motion *indestructible* at the start.[6]

As Mach correctly notes, it was Descartes' intention to *mechanise* or *geometrise* the entire universe.[7] In the beginning, God put a certain amount of motion into the world, and this amount has been preserved ever since.[8] God is the great inaugurator. Having planned, created and energised the universe at the outset, God retired and left it to work like a great clock. Descartes believes that from such first principles he can *deduce* the law of inertia. It is true, as Mach[9] charitably admits, that in a vague sort of way the philosophical speculations of Descartes prepared men's minds for the principle of conservation of energy. But it is impossible to derive what is truly an *empirical* generalisation from general *philosophical* first principles. The Kantian categories, for example, cannot teach us that in Joule's experiment the quantity

$$mgh,$$

where h = a vertical displacement of a mass m
and g = the acceleration associated with the earth's gravitational field, will reappear as the product

$$\text{mass of water} \times \text{rise in temperature}.$$

[5] R. Descartes, op. cit., pp. 287–91.
[6] M., pp. 322–3. The italics appear in the German text. [7] E., p. 28.
[8] R. Descartes, op. cit., pp. 318–29. [9] M., p. 551.

We are nevertheless prepared by Kant to look out for 'something' that must be conserved throughout a change or set of changes.

2. Berkeley and Mach

In the metaphysics of Descartes, there is the table as it is seen, and underlying it—as it were—are the primary geometrical and kinematic qualities of extended body, motion and rest. There is a *seen* order, and also a *hidden* order. In modern molecular theory, the atoms and their motions belong to this hidden order; the unsophisticated modern student of physics or chemistry is usually Cartesian, although he may be unaware of the fact.

Now Berkeley considers that the only table 'which is at all' is the table as seen or felt. Its *esse* is *percipi*:

> For as to what is said of the absolute existence of unthinking things without any relation to their being perceiv'd, that is to me perfectly unintelligible. Their *esse* is *percipi*, nor is it possible that they shou'd have any existence, out of the minds or thinking things which perceive them.[1]

The transition from Descartes to Berkeley is the classical instance in philosophy of the application of Ockham's Razor, the maxim that one 'entity' is preferable to two 'entities'. The table as sense-perceived is enough for Berkeley, and also for Mach.[2]

If the dualism of

mind and body

is essential to the thought of Descartes, the dualism of

percipient and perception

is equally essential to that of Berkeley. Mach takes exception to any kind of dualism; it seems likely however that the ultimate dualism of

knower and known

is necessary to any coherent epistemology, and it is not

[1] G. Berkeley, *The Principles of Human Knowledge*, ed. T. E. Jessop, London: 1949, p. 28. [2] A., pp. 367–8.

surprising that Mach's attempts to avoid even this dualism fail to convince us.

3. The world is our sense-perception

A strange smell near an electrical machine in the year 1785 led to the discovery of ozone. Such is the powerful stimulus of a single sense-perception.

Mach, like the *Gestalt* psychologists, recognises as early as 1885[1] that the so-called 'elements' into which experience is broken down are the product of the analytical activity of the mind, which occurs after the living experience of some kind of organic whole. The analysis of the presented Whole takes place in successive stages as indicated in the diagram, and it is pursued until 'elements' which defy any further analysis are reached. An example of such an element would be 'a sense-perception of green'.

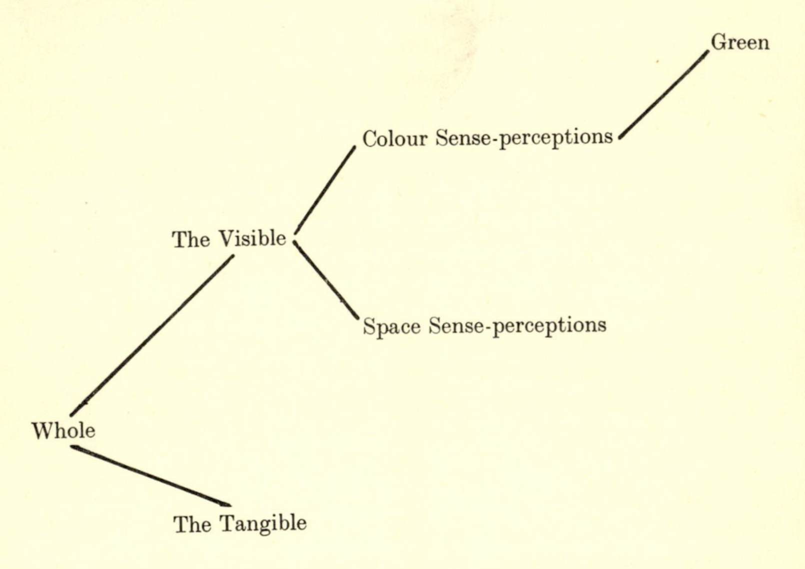

Mach's central assumption must be given in his own words:

> Let us look at the matter without bias. The world consists of colours, sounds, temperatures, pressures, spaces, times and so forth,

[1] A., p. 102.

which *now* we shall not call sensations, nor phenomena, because in either term an arbitrary, one-sided theory is embodied, but simply *elements*. The fixing of the flux of these elements, whether mediately or immediately, is the real object of physical research. As long as, neglecting our own body, we employ ourselves with the interdependence of those groups of elements which, including men and animals, make up *foreign* bodies, we are physicists. For example, we investigate the change of the red colour of a body as produced by a change of illumination. . . . We close our eyes, and the red together with the whole visible world disappears. There exists, thus, in the perspective field of every sense a portion which exercises on all the rest a different and more powerful influence than the rest upon one another. With this, however, all is said. In the light of this remark, we call *all* elements; in so far as we regard them as dependent on this special part (our body), *sensations*. That the world is our sensation, in this sense, cannot be questioned.[2]

And we are wrong if we ask for more than this world which "is our sensation":

One must not attempt to explain a sense-perception. It is something so simple and so fundamental, that the attempt to trace it back to something even simpler, at least at the present time, can never succeed.[3]

Mach is drawing an analogy here between his 'elements' and the common chemical elements; his irreducible sense-perceptions could be called 'atoms of experience'.

In the *Analysis of Sensations* Mach classifies the elements into three sets as shown:[4]

ABC Elements of ordinary bodies, e.g. tables, chairs.
A = a sense-perception unit, e.g. the smell of chlorine, the brownness of the table leg.

KLM Elements of men's bodies.
K = a sense-perception unit, e.g. the blueness of Mach's eye.

$\alpha\beta\gamma$ Elements of volition, memory and the rest, e.g. α = my feeling of joy on climbing Great Gable.

These groups are not however independent. Thus I may think something (α) which makes me blush (K), or which reminds me

[2] P., pp. 208–9. Comma replaced by semicolon, and one additional word italicised. [3] E., p. 44. [4] A., pp. 8–37.

to take an extra chair into the dining-room (A). There is certainly a tendency for certain elements (AB) to remain together, as for example the smell of chlorine (A) which normally abides with the colour of chlorine (B) in what we call 'the substance chlorine'. The stability of the common object, shall we say the kitchen table (ABCD), is however exaggerated. I have only to walk away from it (alteration in KLM) and it becomes smaller visually (abcd). To say that 'it does not really alter' is nonsense, for there is no 'real table' behind the table of experience.

Mach gives ingenious examples of the way in which the elements are all knit together, and concludes:

All elements ABC . . ., KLM . . ., constitute a *single* coherent mass only, in which, when any one element is disturbed, *all* is put in motion; except that a disturbance in KLM . . . has a more extensive and profound action than one in ABC. . . . A magnet in our neighbourhood disturbs the particles of iron near it; a falling boulder shakes the earth; but the severing of a nerve sets in motion the *whole* system of elements. Quite involuntarily does this relation of things suggest the picture of a viscous mass, at certain places (as in the ego) more firmly coherent than in others.[5]

In fact Mach intends 'the one great porridge of experience' to include *all* the elements, i.e. ABC . . ., KLM . . . and $\alpha\beta\gamma$ . . . All is the One Thing, as Anaximander believed:

Thus, perceptions, presentations, volitions, and emotions, in short the whole inner and outer world, are put together, in combinations of varying evanescence and permanence, out of *a small number of homogeneous elements*. Usually, these elements are called *sensations*. But as vestiges of a *one-sided theory* inhere in that term, we prefer to speak simply of elements, as we have already done.[6]

A common body, such as the kitchen table, is not absolutely permanent; it is however a rather stable complex of colour, sound and pressure elements.[7] The ego is a similar relatively permanent complex, but it contains elements of memory, mood and feeling joined to the sense-elements associated with the physical body.[8] The relative permanence of the table and the

[5] A., p. 17.
[6] A., p. 22. Italics added as in German text.
[7] A., pp. 2–3.
[8] Ibid.

ego arises from the fact that these are—so to say—particularly knotty lumps in the porridge of experience. But we are not justified in representing the kitchen chair by the scheme

[ABC]

for if I (my body KLM) do not see it there is no chair, and if I cannot recollect it ($\alpha\beta\gamma$) I am not likely to be able to use it. The term 'kitchen chair' is no more than a 'compendious mental symbol for groups of sensations' and any such symbol has no existence "outside of thought".[9] In the same way there is no final justification for representing the ego as

[KLM].

Just as in classical dynamics every mass point has a field of force which fills all space, and just as the atoms of Boscovich were point centres of force[10] to which no spatial limit of influence could be set, so in Mach's philosophy any common object or any ego must properly extend continuously to fill up the whole 'viscous' mass of the All which is the One. "*Continuity* alone is important". So the "ego must be given up".[11] I can never find the boundary of the tangent galvanometer or any other common 'thing'.

If we like, instead of giving up the ego and the common object, we can extend either or both, so each becomes the super-body or the super-ego, two names for the same thing, the All which is the One.[12] Since there is no detached isolated ego, since there is no detached isolated thing, the picture of physics as 'man observing nature' is radically at fault. For man himself is "a bit of nature".[13] A controlled experiment, in which *events are observed both with and without the observer,* can never be achieved.

According to Mach, the 'world' or 'nature'—and these terms do not refer to something exclusively 'physical'—is ONE. This 'ONE' is a living shifting changing thing; although, as Mach puts it, it is "once there", we may add that 'it is never twice the

[9] P., p. 201. (Lecture given in 1882 *before* the first edition of A., viz. 1886.)
[10] *Roger Joseph Boscovich,* ed. L. L. Whyte, London: 1961, pp. 102–26.
[11] A., p. 24.
[12] E., p. 9. Mach refers to "das *erweiterte* Ich, . . .". [13] P., p. 63.

same', like the river of Heraclitus. The parts of this one thing are called by Mach the 'elements'. The elements include sense-perceptions, which species of element constitutes the subject matter of physics. Since the world is *our* sense-perception Mach has to justify the use of the plural possessive adjective. He would avoid solipsism by the most trite of arguments: since *you* have a body, nervous system and so on, similar to *my* body, nervous system, and so on, then, arguing by analogy, it is probable *you* have a 'consciousness' very like *my* 'consciousness'.[14] Solipsism of the ordinary kind is denied to Mach in any case, because to him no common object and no individual ego enjoys complete integrity; 'all is the one' as Anaximander taught, but Mach converts the proposition to 'each provisionally individual one is the all'.

It is therefore not at all clear as to *who* has the sense-perceptions which make up the subject matter of physics. Mach hopes to avoid the question by his doctrine of "das erweiterte Ich", which is neither more acceptable nor more credible than Berkeley's "some eternal spirit". Since a sense-perception should not necessarily be attached to a specific 'provisional fiction' (my ego), it can be 'unknown' or even 'take a walk' by itself. So Mach at last is driven to deny the kind of dualism which Berkeley consistently maintained, the dualism of perception and percipient.

> Besides, the single sense-perception is neither conscious nor unconscious. It becomes consciously known by being set in the context of present experience. . .
> When the world is sawn and cut up into abstractions, the bits seem so flimsy and unsubstantial that doubts may arise as to whether the world will ever permit itself to be glued together again from them. One might even ask, on occasions, no doubt humorously and ironically, whether a sense-perception or an idea, which belongs to no particular ego, can take a stroll all *alone* in the world?[15]

It is instructive to apply Mach's theory of the elements to the so-called illusions of sense-perception.

> A pencil held in front of us in the air is seen by us as straight; dip it into the water, and we see it crooked. In the latter case we say

[14] A., p. 18. [15] E., pp. 44, 460.

that the pencil *appears* crooked, but is in *reality* straight. But what justifies us in declaring *one* fact rather than *another* to be the reality, and degrading the other to the level of appearance?[16]

The question is admirably posed and quite consistently answered by Mach. Since there is no real pencil behind what appears to sense-perception, all we can say about the simple experiments with the pencil is exhausted by a complete account of the actual sense-perceptions:

When the pencil is entirely in the air, it appears to both the senses of sight and touch as straight.

When the pencil is partly immersed in water, it appears bent to the sense of sight and straight to the sense of touch.

It is in fact impossible for an experience to be genuinely misleading in itself; it is impossible for a sense-perception to be false. As Mach points out, the wildest dream is just as much a fact as any other fact,[17] and if I do see a ghost I do see a ghost.[18] A queer smell occurs; a queer smell is neither true nor false.

This leads Mach[19] to a partial statement of an epistemological position which is a common-place in philosophy, viz. that 'true' and 'false' are predicates applicable to judgements and propositions, but not strictly to sense-perceptions. Suppose one baldly reports that 'the pencil is bent', this is 'error' and 'false' because many of the facts of the given occasion or of the given experience are omitted. The error arises in the suppression of the facts that there is a bowl of water, the fact that the pencil is partly immersed in the water, and that it is straight to the touch. 'The pencil is bent' is in itself a true or correct item, but experience is never a single item. The error arises in the report of an experience, in which report certain items of the experience have been suppressed. Suppose someone says: Everything looks yellow to me. This is error, in the same sense, because he has failed to mention the fact that he has taken a dose of santonin.[20]

Regarded in Mach's way, both 'knowledge' (Erkenntnis) and 'error' (Irrtum) arise out of experience, and indeed out of the

[16] A., p. 10. Italics added as in German text.
[17] A., p. 11. [18] E., p. 116. [19] E., pp. 108–25. [20] O., p. 1.

same experience.[21] There is always a tendency, in reporting experience, to disregard "a whole pile of attendant and relevant circumstances",[22] and this is the source of error. To the extent to which it is borne in mind that the rest of the universe may be relevant to an observation or experiment, to that extent will the report of the experiment be true, a part of genuine knowledge. The last sentence is an approximate affirmation of what has now come to be called the *Mach Principle*.

The illusions of sense-perception must themselves be illusory if the sense-perceptions are our only source of information. If it be metaphysical nonsense to think there is 'anything behind' what appears, it becomes meaningless to affirm that some individual sense-perception is illusory. If on the other hand the reader thinks that it is nonsense *not* to affirm that 'the pencil really is straight', more than likely he is thinking in Cartesian terms. Descartes will permit him to think that the primary qualities of the partly immersed pencil can include the geometrical straight line.

If it be true that science is based on experience and that experience is constituted of sense-perceptions, it is also true that the combined experience of humanity down the ages is entirely inadequate and insufficient to support the great scientific systems which have been built. The experience is not of the right kind, and there is far too little of it; it is deficient both in quality and quantity. Mach has a useful term—*the completion of experience*—which he frequently uses[23] when referring to this kind of question. It will be sufficient here to present two of Mach's own examples. In the first we are asked to consider the slow vibrations of a long elastic rod clamped in a vice at one end.[24] The vibrations can be both *seen* and *felt*. If the rod is shortened, the vibrations become so fast that they can no longer be seen as such; but we abide by the idea of vibrations and, indeed, they can still be felt and recorded graphically. When the rod is shortened still further, it begins to emit a sound which we can *hear*. Again we retain the idea of vibrations and we can still record them by a graphical device,

[21] E., p. 115. [22] E., p. 123. [23] M., p. 588. E., p. 7.
[24] M., pp. 587–8.

or we can calculate their frequencies from a theory which is a kind of extrapolation of what we observed of the more convenient slower vibrations. The experiment can be extended even further. When the pitch of the sound becomes so high that it can no longer be heard we still imagine that the rod is vibrating, and we have good indirect reasons for this belief. Mach draws valuable lessons from this example. In a sense the function of scientific procedures, which include experiment, is to "replace experience".[25] Although the proper province of science must be experience, science is forever 'hastening beyond'[26] experience. Experience, must necessarily be accompanied by thought, and experience without thought would be quite foreign and strange to us.[27] Such thought, whether it be dignified by the grand term of 'theory' or not, must lead to consequences which are verifiable. If neither confirmation nor refutation is possible, the matter has no relevance to science.[28] The experience on which science is properly based must always remain "uncompleted".[29] The thoughts which can interpret for us the largest domain of experience, and which most effectively fill out or complete that experience, are correctly judged the most scientific. Mach considers that the imaginative thought of the investigator is regulated by a *principle of continuity*. This principle he mentions in the same context,[30] but what he means by it is more clearly shown in his discussion of the achievements of Newton. This is the second example of 'the completion of experience', and I think a more significant one:

> This is an achievement of the *imagination*. . . . Newton perceived, with great audacity of thought, and first in the instance of the moon, that this acceleration differed in no substantial respect from the acceleration of gravity so familiar to us. It was probably the principle of continuity, which accomplished so much in Galileo's case, that led him to his discovery. He was wont, and this habit appears to be common to all truly great investigators, to adhere as closely as possible, even in cases presenting altered conditions, to a conception once formed, to preserve the same uniformity in his conceptions that nature teaches us to see in her processes. That which is a property of nature *at any one time* and in any one place, *constantly*

[25] M., p. 586. [26] Ibid. [27] M., p. 587. [28] Ibid. [29] Ibid.
[30] Ibid.

and *everywhere* recurs, though it may not be with the same prominence. . . . [Newton] asks himself: Where lies the limit of this action of terrestrial gravity? Should its action not extend to the moon? With this question the great flight of fancy was taken, . . .[31]

Evidently Mach means by 'the completion of experience' much more than the escape from solipsism discussed previously. The example from Newton's work indicates that he intends rather the entire theoretical part of science. Mach breaks down the distinction between the *concept* of universal gravitational force and a common '*felt*' push or pull. He invites us as it were to think of 'a man in the moon' who can *feel* the force of the earth. There is a certain philosophical advantage in anthropomorphism. The phrase, 'the completion of experience', is characteristic of Mach. By it, we are meant to understand that experience (sense-perceptions) leads on to theory (metrical concepts). Mach's view, that *some of the metrical concepts of physics are refined sense-perceptions*, will be expounded in the next chapter.

4. Phenomenalism

Price has given us a valuable account of that kind of theory of experience which philosophers call 'phenomenalism'. Mach's theory is of this kind. Price begins by telling us that when he sees a tomato, he cannot doubt that "there exists a red patch of a round and somewhat bulgy shape".[1] This is *given* to us, and we therefore call it a 'datum'. We apprehend, or simply 'have', sense-data. Small groups of these sense-data fit together to form single solids. Such a solid, Price calls a family of sense-data.[2] More clearly than Mach, Price urges that the family is *given* with a size and position. The table, for example, may be *given* as being to the left of the chair or behind it, and both table and chair are *given* as *outside* of the observer, in some kind of space. The table, that is some specific or individual family, can be located by two observers who stand apart. It is an "important and puzzling"[3] question to inquire as to whether the space which families occupy is or is not the public space of

[31] M., pp. 228–9. Italics added as in German text.

[1] H. H. Price, *Perception*, London: 1954, p. 3. [2] Ibid., p. 218.

[3] Ibid., p. 246.

classical physics. Price does not resolve this puzzle, but he believes that had we never been given families, we should never have reached the idea of a common physical space.[4] The collected experience of all men is too meagre to add up to a great science like physics. Price, following John Stuart Mill, is therefore driven to the idea of 'possible sense-data', and he suggests that the family "is really a persistent whole of *possibilities*".[5] This is preferable to Mach's sense-perception which takes a stroll on its own. Price notes that such theories of experience, which include Mach's theory and other kinds of phenomenalism, are generally on the side of common-sense, for they do not require us to believe in anything "which we do not already and in any case know to exist".[6] Such theories also respect Ockham's Razor, for they do not require us to multiply entities unnecessarily.

Although Mach's phenomenalism is made more acceptable by Price's additions, there remain some serious difficulties. Two of these were discussed many years ago by the Italian philosopher, A. Aliotta. Mach was led by his studies in classical thermodynamics to the view that an entirely *mechanical* interpretation of nature is not possible.[7] It would be an error to think of atoms and molecules as real things, and Mach does not commit the error; but he makes *the same kind of error* by raising the sensorial elements in the web of experience to the rank of reality:

> When, however, he [Mach] endeavours to build up a new intuition of the world on the ruins of the mechanical theory, and substitutes the element of sensation for the material atom, he does but replace mechanical by sensorial mythology. The atom was the hypostasis of an abstraction; what else is the sensorial element?[8]

According to Aliotta, Mach substitutes a 'psychological atomism' for a 'mechanical or physical atomism'. Hume himself commits the same fallacy.[9] The effect of psychological atomism

[4] Ibid., p. 252. [5] Ibid., p. 262. [6] Ibid., p. 282.

[7] A. Aliotta, *The Idealistic Reaction Against Science*, trans. Agnes McCaskill, London: 1914, p. 65.

[8] Ibid., pp. 65–6.

[9] D. Hume, *Hume's Enquiries*, ed. L. A. Selby-Bigge, Oxford: 1902, p. 19.

in Mach is heightened by his use of the symbols ABC . . ., KLM . . . and $\alpha\beta\gamma$ . . .

Aliotta's second adverse criticism of Mach is more radical; it is directed towards Mach's doctrine of the ego:

> The *Ego* is not a content of experience which simply stands in relation to other contents: its special characteristic is that it knows that which stands in relation to itself, and is conscious of its own relations, whereas the opposing member knows nothing of the *Ego*.[10]

The ego cannot possibly swell out and become coextensive with *everything* that is experienced or known if, at the same time, the ego is to remain that which knows or experiences *something* or *anything*.[11]

5. Hooke and Mach

The point of departure is sense-perception. The outward journey into physics is continued through the concepts. Were Mach alive now he would think, as I do, that some modern writers have too greatly stressed the freedom of the human mind when engaged in constructing concepts and conceptual systems. Since the deepest truth about the abstract concepts is that they "draw their ultimate power from sensuous sources",[1] the creator of the concept is certainly not *absolutely* free; for he is constrained by the whole experience of the facts—past or present, others' or his own.

> The concepts of modern physics [1901] will stand comparison, in point of precision and height of abstraction, with those of any other science; but they offer at the same time the advantage that they can always be traced back with ease and certainty to the sensational elements on which they are built up. For science the gulf between intuitional presentation and conceptual thought is not so great, and is not unbridgeable.[2]

Mach's work as a whole is a criticism of the concepts of physics. Such criticism is a good way to write a philosophy of physics. It was certainly the best way for him.

[10] A. Aliotta, *The Idealistic Reaction against Science*, p. 84.
[11] H. Dingle, *Through Science to Philosophy*, Oxford: 1937, pp. 344–5.
[1] E., p. 387. [2] A., pp. 363–4.

The outward journey from sense-perception to the metrical concepts must be followed by a return journey—the return for more experience to verify or falsify the intellectual constructions. Just as life flows up from the ground of experience into the living intellectual tree, so the soundness and well-being of that tree is demonstrated by its power to throw out new roots, return thus to experience, experience which has become wider and deeper than before. Mach uses the verb 'return' (wiederkehren): "So all intellectual life proceeds out from sense-perceptions, and returns back again to them."[3] It is an important doctrine. The mind must *return* from its conceptual exercises to the ground of new experience. The philosophy of return is no mere item in a scientific methodology, tamely listed as 'verification'; it requires a lively metaphor, which Robert Hooke found first in 1665:

> So many are the *links*, upon which the true Philosophy depends, of which, if any one be *loose*, or *weak*, the whole *chain* is in danger of being dissolv'd; it is to *begin* with the Hands and Eyes, and to *proceed* on through the Memory, to be *continued* by the Reason; nor is it to stop there, but to *come about* to the Hands and Eyes again, and so, by a *continual passage round* from one Faculty to another, it is to be maintained in life and strength, as much as the body of man is by the *circulation* of the blood through the several parts of the body, the Arms, the Feet, the Lungs, the Heart, and the Head.[4]

I conclude with an example which requires the theoretical mechanics of Huygens and Newton. Suppose one has a spiral spring and finds that a standard 100 g mass suspended from it extends the length of the spring by 5 cm. This means that a force of 100×981 dynes extends the spring 5 cm. By experiments with other standard 100 g masses it is found that the spring 'obeys' Hooke's law. The unstretched length of the spring is 50 cm. It is now found that a lump of lead A stretches the spring by 15 cm, and so the deductions

$$\text{weight of A} = 300\text{ g wt.},$$
$$\text{mass of A} = 300\text{ g}$$

are drawn. A is now whirled round in a horizontal circle on the

[3] E., p. 144.

[4] R. Hooke, *Extracts from Micrographia*, Alembic Club Reprint no. 5, Edinburgh: 1902, p. 12.

floor of the room at a steady speed. When the speed is 150 cm per second, the spring is stretched to the 'radius' 56·1 cm or by an extension of 6·1 cm. By Hooke's law and Newton's second 'law' the 'central force' is

$$\frac{6{\cdot}1}{5} \times 100 \times 981 \text{ dynes.}$$

From Huygens' formula for central acceleration and again from Newton's second 'law', this force must also be

$$\frac{m \times 150^2}{56{\cdot}1} \text{ dynes}$$

where m is the mass of A in grammes. Equation of the two expressions for the central force gives $m = 298$ g, in good agreement with the value obtained more directly in the 'static' experiment. On the hand, A *feels* about as heavy as three of the standard 100 g masses. Also when 'weighed' on a balance its mass is found to be 301 g. In teaching mechanics, I have often done this experiment, but the experimental error has been up to 5 per cent. The example, although not a complicated one, presents more than a single outward journey and a single return journey *from* and *to* experience. Hooke's metaphor of a circulation is required; the doctrine of return should be depicted as

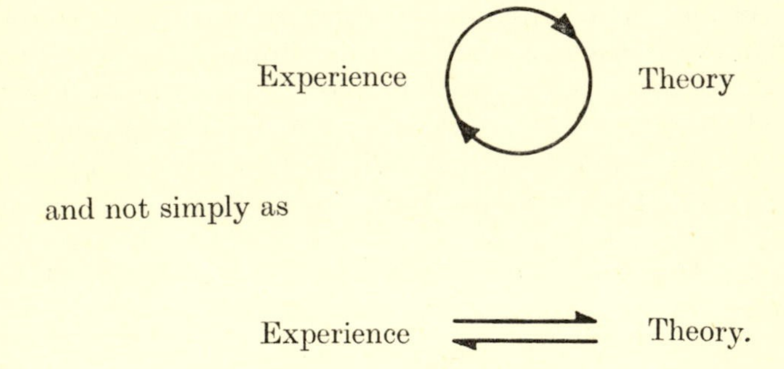

6. The one realm of science

Reference has been made to two kinds (Descartes, Berkeley) of philosophy of nature. A sophisticated modern form of the

Cartesian two-realm philosophy has been sketched out by C. G. Hempel:

A scientific theory might therefore be likened to a complex spatial network: Its terms are represented by the knots, while the threads connecting the latter correspond, in part, to the definitions and, in part, to the fundamental and derivative hypotheses included in the theory. The whole system floats, as it were, above the plane of observation and is anchored to it by rules of interpretation. These might be viewed as strings which are not part of the network but link certain parts of the latter with specific places in the plane of observation.[1]

Mach hopes to give an intelligible account of physics without two orders of substance (Descartes) and without two planes (Hempel). The realm of science is one, an integrated whole. The possibility of Mach's success is seen more easily in the terms of Hempel than in the terms of Descartes. All the concepts of physics are derived from sensuous sources; Hempel's two planes are at once linked. For the concepts belong to the spatial network of scientific theory, whereas the sensuous sources spring from observation. Mach has not merely identified, located or described Hempel's strings; he fuses the two planes together by proving that there is natural continuity between sense-perception and intellectual conception and the strings are not needed. Theory does not float above observation; theory is *in* science like yeast in good bread. This is the main part of Mach's programme for the philosophy of science; his success or failure therein may be judged by the reader from what follows.

The British philosopher G. Dawes Hicks considered that conceptual thinking and sense-apprehension are like opposite ends of the same spectrum:

. . . the process of conceptual thinking is . . ., in truth, a development from the more primitive process of sense-apprehension and continuous with it in nature, . . .[2]

He therefore supports Mach's view that the concepts of physics

[1] R. Carnap, *Philosophical Foundations of Physics*, New York: 1966, p. 266. Quotation from C. G. Hempel, 1952. I am indebted to Dr Mary Hesse for directing me to Hempel's writings.

[2] G. Dawes Hicks, *Critical Realism*, London: 1938, p. 154.

derive their power from sensuous sources. If no man anywhere or at any time had ever *felt* a push or pull, no one would have been constrained to create a metrical concept of force. We hardly think, perhaps, of a 'man in the sun' who feels the tug of the planet Jupiter. Yet:

> . . . there is no incongruity in describing "forces" as the sort of entities which we become aware of in and through our perception or feeling of strain or effort, nor in asserting that physical bodies, such as the sun, are influenced by and exert "force" in this sense. In other words, there is no more reason for identifying tension or strain with the consciousness of it than for identifying colours or sounds with the consciousness of those colours or sounds.[3]

7. Examples leading to Mach's theory of the metrical concepts

Mach's popular lecture 'On the Fundamental Concepts of Electrostatics'[1] given at Vienna in 1883 is one of his more important writings. Mach informed his listeners that, in order to acquire a "lucid and easy grasp of the phenomena",[2] they must learn how to reproduce "the facts accurately in thought",[3] and that the means to this end are "the *metrical* concepts"[4] (die Massbegriffe) of electricity.

When a glass rod rubbed with silk is allowed to touch each in turn of two small pith balls, suspended from the same point by dry silk threads, the two balls repel one another as shown in Fig. 1. The force of repulsion between the balls is an ordinary

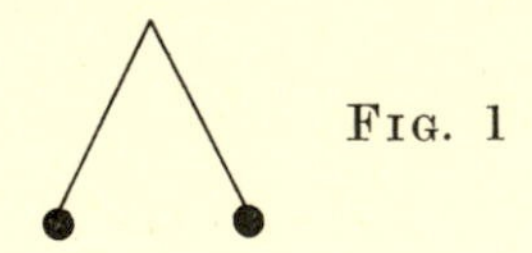

FIG. 1

mechanical one, and it does in fact do work on the balls against the force of gravity on them. By delicate experiments with the torsion-balance, Coulomb proved that if the distance between the balls was doubled, the force became four times less; if the

[3] Ibid., p. 179.

[1] P., pp. 107–36. [2] P., p. 107. [3] Ibid. [4] Ibid.

distance was trebled the force was nine times less, and so on. The magnitude of the force was determined by the distance, according to an inverse square law.

Now suppose (see Fig. 2) that bodies A and K have both been touched with the rubbed rod, that they are 2 cm apart and that the force of repulsion between them is 1 mg weight. It is also to be supposed that this mechanical force of repulsion can be measured. A is now touched by an exactly equal body B. Force measurements show now that the force between A alone and K is $\frac{1}{2}$ mg weight, that the force between B alone and K is also $\frac{1}{2}$ mg weight, but that the total force between A and B together and K is 1 mg weight as before.

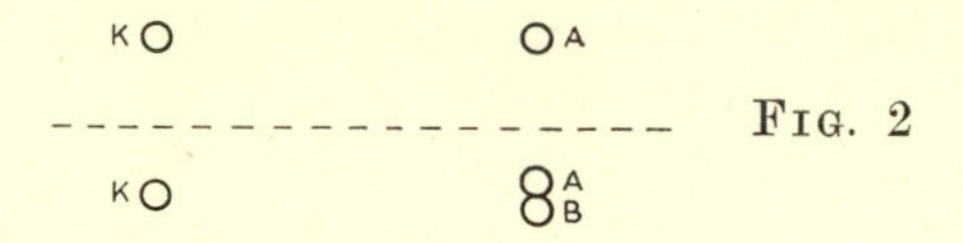

FIG. 2

The way we proceed to interpret these facts is, as Mach says, convenient, but not necessary. We imagine an electrical fluid originally all in A, but later shared equally with B, and we imagine the force to be proportional to the amount of the fluid.

Hence, *the divisibility of electrical force* among bodies in contact *is a fact*. It is a useful, but by no means a necessary *supplement to this fact*, to imagine an electrical *fluid* present in the body A, with the quantity of which the electrical force varies, and half of which flows over to B. For, in the place of the new physical picture, thus, an old, familiar one is substituted, which moves spontaneously in its wonted courses.[5]

It was perhaps the great prestige of Newton which persuaded Coulomb to adopt a mode of interpretation, leading to an electrical law in form identical with the law of gravitation, and in which law 'quantity of electricity' is a formal analogue of 'mass'.[6]

Although it is useful to imagine electrical fluids, it would be misleading to look "in nature for the two hypothetical fluids",

[5] P., p. 110. Italics added as in German text.
[6] P., p. 173.

which in fact we have added to our experience as "simple mental adjuncts".[7] For it is the forces, not the fluids, which are factual:

> . . . and the most proper course is always, after the general survey is obtained, to go back to the actual facts, to the electrical forces.[8]

I think Mach should have added here that the so-called electrical forces are qualitatively identical with any other mechanical forces. They are not electrical in themselves; they are proper mechanical forces whose principal effect is to accelerate a mass.

The main point of this discussion is strengthened by what follows on the electrolysis of acidulated water.[9] Suppose we imagine water as columns of hydrogen and oxygen fitting in to each other. The chemical decomposition of water is the drawing apart of the two columns in opposite directions. As the decomposition proceeds, a quantity of electricity proportional to the mass of water decomposed passes in one direction across the voltameter; or, if preferred, the *two* electrical fluids can pass in *opposite* directions, one quantity proportional to the mass of hydrogen set free and the other proportional to the mass of oxygen set free. The first two laws of electrolysis discovered by Faraday are seen by Mach as an extension of the Coulomb analogy between 'mass' and 'quantity of electricity'. In the electrolysis there are *currents of mass* as well as *currents of electrical quantity*. If it were possible to classify the metrical concepts of physics one would expect to find 'mass' and 'quantity of electricity' in the same class; 'force' would be in a different class.

The example which follows is not given by Mach, but I think he would have approved of it. Suppose 100 g of water at 10°C is mixed with 100 g of water at 30°C. If the experiment is done carefully in a thermos flask, the temperature of the mixture is 20°C or nearly so. The two equal masses of water correspond to the two equal bodies A and B of the previous example; 'temperature' in the present example corresponds to 'force' in the previous one. With appropriate alterations Mach's commentary on the previous experiment can be applied again. *The*

[7] P., p. 113. [8] P., p. 114. [9] P., p. 118.

divisibility of temperature between the two sets of water *is a fact* which occurs on mixing them:

$$\frac{10 + 30}{2} = 20.$$

It can be imagined that the warmer water contains a *thermal fluid*, called *heat*, and that this heat passing from the warmer to the colder water is divided equally between the two. If so, it must also be imagined that the temperature of a body is proportional to the amount of heat fluid in it. But the *facts* of the matter concern temperature, not heat. It would be a waste of time to seek for the heat fluid in nature. The concept of heat fluid is an *intellectual addition* (Zutat), and is not necessarily required by the phenomena of the experiment. In the science of heat, one should return to the actual facts, i.e. the temperatures.

Again, if it were possible to classify the metrical concepts of physics, we would expect to find 'mass', 'quantity of electricity' and 'heat' together in one class; 'force' and 'temperature' would be together in a different class.

2 Metrical concepts

1. The ordinal test

Although all the metrical concepts of physical science draw their power from sensuous sources, the character of the relationship between sense-perception and concept may vary radically. This is the plain meaning of Mach's distinction between 'force' which is a matter of fact (eine Tatsache), and 'quantity of electricity' which is "a supplement to this fact" (eine Zutat). It is possible to develop this distinction of Mach,[1] and in the present work I have attempted to derive from it a classification of the metrical concepts.

Let us take as initial examples the five metrical concepts: force, temperature, mass, heat and quantity of electricity. Specific values of these types of quantity are derived from experiments which may include measurements and calculations. There is however a certain difference in the relationship between these concepts and the experience, some of it qualitative, which gives rise to them. 'Behind' the metrical concepts of 'force' and 'temperature' there lies a group of simple acts of sense-perception, which can, under favourable circumstances and within a rather narrow range of experience, be *ordered* by the percipient according to *felt* intensity, in an order which tallies with the metrical values subsequently reached by operations and measurements. The use of 'felt' as an alternative to 'sense-perceived' is found in Hume.[2]

This notion of *ordinal correspondence* is best understood by two simple examples. Suppose there are three baskets of

[1] J. Bradley, *The Ordering of the Concepts of Classical Physics and Chemistry*, M.Sc. Dissertation, University of London, 1951, pp. 71–85.

[2] D. Hume, *Hume's Enquiries*, p. 62.

potatoes, called A, B and C. I lift them and report:

A is very heavy,
B is light,
and C is fairly heavy.

By almost immediate sense-perception I can *order* the three baskets as

A C B

putting the most heavy first. Now suppose I measure the weights of the baskets using a spring-balance, which is in fact a force-meter. The three measurements are:

A 28 lb. wt.,
B 5 lb. wt.,
and C 15 lb. wt.,

from which I can derive *the same order*

A C B.

For the second example, let A be the Bunsen burner flame, B boiling water, and C boiling acetaldehyde. A is so hot it burns and chars my hand badly, B brings up blisters but does not char my hand, and I can put my hand into C without great discomfort. The order of felt hotness is therefore

A B C

with the hottest first. If now the three temperatures are measured by appropriate thermometers it is found:

A 850°C
B 100°C
C 21°C

and the order is the same, viz.

A B C

with the largest number placed first.

There is then an *ordinal correspondence* between intensities of *felt* (*immediate* and almost *unreasoning*) experiences, and a group of experimental (instrumental) measurements. *Behind*, as it were, the *metrical concepts* of force and temperature, are *shadow concepts*, or *pre-concepts* which are in fact *sense-perceptions*; and between these *metrical concepts* and the *pre-concepts* there is ordinal correspondence. It is no doubt bold to describe a sense-perception by a term which includes the word concept, and Mach never did this. But the procedure finds a philosophical defence in the writings of Dawes Hicks; it is a sufficiently loud proclamation of the truth that *the metrical concepts of physics are not absolutely free creations of the human mind.*

Expressing the matter less carefully 'force' and 'temperature' can not only be measured; they can also be felt. It is evident to anyone with a sufficiently clear understanding of classical physics that the same is not true of 'mass'.[3] For example, suppose I am confronted by three golden spheres A, B and C. A is a very large one, B a medium size and C is a little one. I may judge that in order of decreasing mass the spheres should be ordered

$$\mathrm{A} \quad \mathrm{B} \quad \mathrm{C}$$

but I cannot do so correctly unless the spheres are known to be solid and homogeneous. If this is known to be the case I can legitimately order the volumes of the spheres as

$$V_{\mathrm{A}} \quad V_{\mathrm{B}} \quad V_{\mathrm{C}}$$

and this is based on acts of immediate sense-perception. Even so I can only *mediately* judge the mass order as

$$m_{\mathrm{A}} \quad m_{\mathrm{B}} \quad m_{\mathrm{C}},$$

from the knowledge that the mass of a homogeneous material is proportional to its volume. Again, I try to lift the spheres. I find I cannot lift A, B is very heavy, C is light. I may correctly judge there will be an ordinal correspondence between the three trials and the weights $m_{\mathrm{A}}g$, $m_{\mathrm{B}}g$, and $m_{\mathrm{C}}g$. From Newton's second 'law' and the knowledge that 'g' is everywhere the same in the neighbourhood I can infer the mass order, but the inference is far indeed from being immediate. There is no way out. If I take the three spheres with me out into remote space, my failure is still complete. I push each of the spheres away

[3] J. Bradley, "Discussion of Professor F. A. Paneth's Second Article," *British Journal for the Philosophy of Science*, XIV, no. 53, 1963, p. 39.

from my body with a constant force P. I may judge correctly, with respect to the three relative accelerations away from my body

$$f_A, f_B, f_C,$$

that

$$f_A < f_B < f_C.$$

To infer from these inequalities that

$$m_A > m_B > m_C$$

I require to be well-versed in Newtonian dynamics and algebraic operations:

if $\qquad P = m_A f_A = m_B f_B = m_C f_C,$

and if $\qquad f_A < f_B < f_C,$

it follows that $\qquad m_A > m_B > m_C,$

but this is evidently quite remote and mediate. The metrical concept 'mass' is absolutely imperceptible; it has no *direct* relationship whatever to experience or sense-perception. The same is true of 'quantity of electricity' and of 'heat'. Mach's point that, nevertheless, such concepts can be traced back to experience, remains valid; we are constrained by our experience eventually to make and to use such concepts.

I shall now try to state all this in the form of an ordinal test which is satisfied by 'force' and 'temperature' but not by 'quantity of electricity', 'heat' and 'mass'. Suppose values of a certain metrical concept are represented by the symbol C. Such values are reached generally by a process of refinement from experience and sense-perception. The process of refinement can include selection or analysis, experimental operations, measurements, calculations and the creation of concepts other than the one under discussion; such a process may be variously short or long, simple or elaborate. Suppose now this process be completed in three examples, the three values C_1 C_2 C_3 reached, and that the order of these values is

$$C_1 \quad C_2 \quad C_3,$$

not necessarily distinguished from

$$C_3 \quad C_2 \quad C_1.$$

Suppose further that a percipient, by one or more acts of sense-perception applied to each of the examples before the processes of refinement are carried out, and without making use of knowledge beyond the given information of the one or more acts of sense-perception, can judge that, when the processes of refinement have been carried out, the found order will be

$$C_1 \quad C_2 \quad C_3,$$

not necessarily distinguished from

$$C_3 \quad C_2 \quad C_1;$$

then, even if this correspondence is achieved only in some favourable examples, and by no means universally, the metrical concept whose values are C is said to have passed the ordinal test. If the concept is a vector quantity, the symbols C_1 C_2 C_3 refer to the scalar coefficients (quantities) and not to the directed unit.

The statement that the order

$$C_1 \quad C_2 \quad C_3$$

is not to be distinguished from the order

$$C_3 \quad C_2 \quad C_1$$

needs clarification. Returning to the second of the simple preliminary examples, the one concerning the Bunsen flame, boiling water and boiling acetaldehyde, there is no doubt that the percipient *feels* the Bunsen flame as the hottest; there is some doubt whether he feels this warmth-circumstance (Mach uses the word 'Wärmezustand') as *more* than the warmth-circumstances of the two boiling liquids. If asked to put the 'felt hotnesses' in order, *more* or *less*, he may not know whether the order should be

Bunsen flame   Boiling water   Boiling acetaldehyde

or

Boiling acetaldehyde   Boiling water   Bunsen flame.

On the other hand if the order

$$C_1 \quad C_2 \quad C_3$$

is not distinguished from

$$C_3 \quad C_2 \quad C_1,$$

temperature will certainly 'pass the ordinal test'. To keep this mode of classification within the spirit of Mach's ideas it is essential that 'temperature' should be classified with 'force'. It is a curious fact noted by Mach,[3] that when Celsius devised the centigrade mercury thermometer in 1742, he called the ice/water point 100° and the boiling water point 0°. This is an interesting proof, if one were needed, that 'hot' is not necessarily regarded as *more than* 'cold'. Strömer reversed the numbers of the Celsius scale later.

It is convenient to call those concepts which pass the ordinal test 'O' concepts, and to call those which do not pass the test 'non-O' concepts.

2. The inaugurating concepts

Each one of four main branches of physics is 'set off', as it were, by an individual 'O' concept, or by the 'shadow concept' which 'lies behind' this individual 'O' concept. The concepts of force, temperature, intensity of illumination and frequency (i.e. frequency of longitudinal air waves) are inaugurating concepts in this sense.

The deep root of dynamics is the sense-perception of pushes and pulls.[1] This is one of Mach's earliest insights (1871) and he frequently returns to it. It may be possible to give a logical and complete account of the science of mechanics without employing the concept 'force' at all. But to do so is to treat the historical evolution of the science with disrespect.[2] The simple fact is that *there are forces* in nature and in our laboratories:

> In the case of a piece of iron lying at rest on a table, both the forces in equilibrium, the weight of the iron and the elasticity of the table, are very easily demonstrable.[3]

And what is there has priority over what is thought.

Just as the deep root of mechanics is the sense-perception of pushes and pulls, so the deep root of the science of heat and thermodynamics is the sense-perception of cold, cool, lukewarm and hot.[4] The metrical concept 'temperature' and the metrical

[3] W., p. 12.

[1] GG., p. 32. [2] M., pp. 317–24. [3] M., p. 319. [4] W., p. 3.

concept 'force' are both 'O' concepts. Moreover the late P. W. Bridgman is correct in asserting that it is temperature which "sets thermodynamics off".[5] In the terms of this essay, 'force' and 'temperature' are inaugurating concepts.

In optics the concept 'intensity of illumination' bears the same kind of relationship to the concept 'quantity of light' as, in heat, the concept 'temperature' bears to the concept 'quantity of heat'. The former member of each pair is an 'O' concept and also an inaugurating concept. Mach admires Lambert because he "bases everything on . . . intensity of illumination, a thing that can be observed".[6]

Mach is interested in our sense-perception of 'musical pitch' which stimulates us towards the study of sound. Evidently the sense-perception of pitch and the metrical concept of frequency (of longitudinal waves) are in ordinal correspondence. The example is a good one, and the need for three values of the metrical concept is very clear in this case. Mach's words suggest the ordinal test:

> Further, we not only *distinguish* between tones, but we also *order* them in a *series*. Of three tones of different pitch, we recognise the middle one immediately as such. . . .[7]

There are thus four major branches of physics—mechanics, heat, light, acoustics—each with its respective inaugurating concept—force, temperature, intensity of illumination, frequency. Mach indeed would add yet a fifth branch of physics, geometry itself. If it is admitted that geometry is a science, it is easy to see that 'length' is its proper inaugurating concept. This bold conception of Mach's will be the main part of the next chapter.

Has the science of magnetism and electricity an inaugurating concept? I think not, and Mach seems to agree.[8] Further, there is no specifically 'electrical' 'O' concept. There is a true difference therefore between electricity and any other branch of physics. If one wished to maintain that there is a valid preference for a mechanical model of electrical phenomena, rather than an

[5] P. W. Bridgman, *The Logic of Modern Physics*, New York: 1960, p. 117.
[6] O., p. 16. [7] A., p. 275. [8] M., pp. 597–9.

electrical model of mechanical phenomena, I think the case could be well made along these lines. It is interesting that in the comparatively modern M K S system of electrical units, the definitions depend on the mechanical force between linear conductors. Our electrical experiences can be broken up into sense-perceptions: these sense-perceptions are organically related to 'O' concepts of *other* branches of physics (e.g. force and temperature): there is however no specific electrical sense-perception, and therefore there is neither electrical 'O' concept nor inaugurating concept for magnetism and electricity. More simply, *there are no electrical phenomena.*[9]

3. An inaugurating 'O' concept: Temperature

Mach's account of the concept of temperature shows him at his finest. It is elaborate and extended; for convenience this section is divided into sub-sections, each with its own sub-heading.

(a) *'Felt-warmth', 'warmth-condition', temperature*

Mach distinguishes between the sense-perceptions of warmth and cold, which, he agrees, are certainly the point of departure of the subject of heat;[1] the warmth-condition considered as obtaining or even existing[2] in nature prior to any possibility or enterprise of measurement; and the metrical concepts of temperature, of which there are many. The German words for these three notions are *Wärme-empfindung*, *Wärmezustand* and *Temperatur*.

Mach's emphasis on the fact that there are at least two orders to be considered, viz. the order of temperature proper and the order of felt-warmth or warmth-condition, is of the greatest value. His further distinction between felt-warmth and warmth-condition is also worth keeping; the former may be taken merely to cover the small range of human experience, whereas the latter is at least a large extrapolation beyond this. Indeed,

[9] The reader may take this as an academic joke. If he insists that electric shock is a phenomenon, I ask him, with what metrical concept of electricity is it in ordinal correspondence?

[1] W., p. 3. [2] W., p. 48.

Mach uses the existential phrase 'es gibt' in his affirmation that *there are* warmth-conditions in nature:

> There are warmth-conditions in nature, but the concept of temperature exists only through our arbitrary *definition*, which might of course have been contrived otherwise.[3]

The order of warmth-condition 'lies behind'—so to say—the order of temperature. Felt-warmth and warmth-condition are the shadow-concept or preconcept of temperature. The metrical concept of temperature 'passes' the ordinal test and is thus an O' concept.

(b) *Temperature not a vector*

Mach is concerned to emphasise the analogy between force and temperature, both 'O' concepts with shadow concepts 'behind' them:

Concept	*Shadow-Concept*
Force	Push, pull
Temperature	Felt-warmth (with warmth-condition).

We might say that temperature begins its intellectual life as a vector quantity: felt-warmth 'comes vectored'—so to say—, as a man standing near the fire may feel 'warm on one side'. Before the concept of temperature is developed, the vector quality of felt-warmth is refined out and removed from it. In classical mechanics, force *remains* a vector, and the composition of two force vectors by the parallelogram law is a part of the theory of statics.

(c) *Temperature and volume*

A familiar experience, that the volume of a solid, liquid or gas changes when it is heated, is the basis of thermometry. To use this change in volume as an index of the change in warmth-condition is, *prima facie*, to replace one kind of sense-perception by another kind. It is much more difficult however for the sense of sight to be deceived by a change in volume, particularly when this is displayed on a numbered scale, than it is for the sense of touch to be deceived by a feeling of warmth or cold.

[3] W., p. 48.

Moreover, different observers are apt to agree about a reading on a scale, and at least a degree of communication becomes possible. With an appropriate device, a very small change of volume can be noticed; the corresponding small change in warmth-condition would be missed.

Giving Hume's words a new context, there is no necessary connection between temperature and volume. The decision to measure the former in terms of the latter can only be justified *a posteriori* or pragmatically. If a man had no sense-perception of warmth or cold he could be seriously misled by the use of a water volume thermometer or by the use of a gas thermometer. With the former, what are in fact two different warmth-conditions could be misinterpreted as one identical temperature;[4] a standard way of using the latter is to permit no variation of volume at all and to estimate a change in temperature through a change in pressure. An uninformed person who naïvely associated volume and temperature could in fact make no reliable use of a thermometer at all. One has to know a little about what one is doing. For example, it would be rather stupid to think the extension of a metal wire, brought about by a mechanical load, is a reliable indication of a rise in temperature.[5]

Mach's acute analysis of the connection between volume and temperature shows that the mode of connection of such things is arbitrary; the choice of what things to connect is also arbitrary.

(d) *A continuum*

As Mach says, a continuum is a manifold of a property A such that, between two values of A which exhibit a finite difference, there is no finite number of values of A. The idea of a continuum is mathematical. In a strict sense, there is no such thing as a physical continuum. If for example there were a continuum of warmth-conditions, there would have to be an infinity of warmth-conditions even within the restricted range of felt-warmth. Such an infinite set could not be 'given'; infinity is not a number in classical arithmetic, and no one could receive

[4] W., p. 43. [5] W., pp. 39–40.

the gift. Logically, in the same way, there is no 'given' continuum of 'seen length'. When the graduated list is fixed to the thermometer, neither the 'felt' warmth-conditions nor the 'seen' lengths are truly given continua. They cannot be 'given'. There is, as it were, the replacement of one false continuum by another false continuum.[6] It is however, as Mach puts it, "a convenient fiction which does no harm, to regard the system of warmth-conditions as a continuum".[7] The point is that our experience does not oblige us to reject the hypothesis that the manifolds of volumes and warmth-conditions are continua. We do not seem to 'come across' an intermediate volume which a gas cannot have; nor do we find some degree of felt-warmth forbidden to us. Our positive faith is that of a double negative; our experience does *not* require us to *dis*believe in the order of warmth-conditions as a continuum.

(e) *Temperature as 'level' or 'pressure' of heat*

That remarkable genius, Joseph Black, regarded temperature as the general level of heat and, in another place, he refers to the equilibrium of heat.[8] To Black, heat itself was an indestructible fluid. A room without animals or plants, but containing a variety of chairs, tables and other inanimate objects, to Joseph Black is like a series of *connected* vessels, the vessels being a variety of shapes and sizes, but all containing the heat fluid at the same level, like the water in Pascal's vases. This level Black takes as the temperature; more precisely, it is the warmth-condition. Black had a clear understanding, in the language of the eighteenth century, that temperature is 'sensible' whereas heat is not 'sensible'. This is the essential distinction between an 'O' concept and a 'non-O' concept. Temperature is not a quantity like volume; it is a potential like fluid pressure or electrical pressure. Mach occasionally uses the French word 'niveau' to denote temperature or warmth-condition.

The *concept of temperature* is a *level-concept* like the height of a heavy body, the velocity of a moving body, the electrical potential, the magnetic potential or the chemical potential. Thermal flow takes place between two bodies at different temperatures in the same

[6] W., p. 72. [7] W., p. 77. [8] W., p. 156, quotation from Black.

way as a current of electricity occurs between two bodies at different electrical potentials.[9]

This being so, it is curious that it has been decided to measure temperature, *which is not a quantity*, in terms of volume, *which is a quantity*. The numbers on the thermometer scale are not numbers which are used to count out 'so much' temperature, for that would be nonsense. In principle they could, perhaps should, be replaced, by a large set of 'fixed points', the marks for melting ice, melting wax, melting camphor, melting naphthalene, boiling water and so on. If the numbers of the temperature scale are compared with the numbers of the dwelling-houses in a road, the desired operation in the analogical case would be to take down the numbers and call the houses 'The Villa', 'Mon Repos', 'Sea View' and so on. An increment ΔV of volume of thermometric substance may be interpreted conventionally as a 1° rise in temperature. A further rise of 1° may be associated with a further volume increase of ΔV. There is no objection whatever to writing

$$\Delta V + \Delta V = 2\Delta V,$$

but one must beware of thinking that temperatures can be counted in this same way. The numbers on the temperature scale are

. . . merely *signs* of the warmth-condition; they are not equal parts —they cannot be counted—of the *general* characteristic called the warmth-condition itself.[10]

Mach inquires how and why Black came to think of the inanimate objects in a room as a variety of vessels filled up to the same constant level with what Black himself called 'the matter of heat'. Characteristically he finds that in the first place the idea springs from crude observations.[11] If a hot object touches a cold object of the same material, or if they are left either near each other or in contact, soon they begin to feel the same. Their felt-warmths equalise. This simple fact in itself suggests that the thermometer, although it must in the first

[9] W., p. 57. [10] W., pp. 69–70. [11] W., p. 39.

place register its own temperature, registers also the temperature of that in which it is immersed. This is assumed in every measurement of temperature.

If the two objects in contact are, for example, wood and iron, they do not feel alike even when they have been left a long time. For this and other reasons, 'seen volume' is substituted for 'felt-warmth'. It is better therefore to define 'equality of warmth-condition' before one defines 'warmth-condition' or 'temperature' itself.

Disregarding mechanical, electrical and other forces, *warmth-conditions of different bodies shall be taken as equal when these different bodies determine in each other no volume changes.*[12]

Suppose by this criterion it is found that

Warmth-condition of A = warmth-condition of C,

and that

Warmth-condition of B = warmth-condition of C,

it *does not follow logically* that

Warmth-condition of A = warmth-condition of B.[13]

Further experience is required in this case. If A and C effect in each other no volume changes, and also if B and C effect in each other no volume changes, then *it is in fact found experientially* that A and B also effect in each other no volume changes. Temperature itself must in the end be defined in such a way that there is a one-to-one correspondence between the orders of warmth-condition and temperature; the above statement applies therefore to temperature as well as to warmth-condition.

(f) *Temperature scales*

Mach gives an excellent account[14] of the history of scales of temperature. Only a few important points are noted here.

Amontons was the first physicist to suggest that temperature changes might be indicated by the pressure changes of a gas held at constant volume. He also knew that two coexistent phases represent a 'fixed point' of warmth-condition, and so of temperature. The possibility of an absolute zero of temperature,

[12] W., p. 41. [13] W., p. 41. [14] W., pp. 6–15.

i.e. the temperature at which the pressure of a constant volume of gas would become zero, was recognised by Amontons.

The important facts concerning the pressure, volume and temperature of gas discovered by Boyle, Mariotte, Charles, Amagat, Regnault and others led eventually to the modern precise gas thermometers.

Gay-Lussac used a mercury thermometer in order to study the expansion of various gases. He found in 1802 that all gases expand to the same extent, which is a most remarkable fact. Thus between the two fixed points of warmth-condition, marked as 0° and 100° on the mercury-in-glass scale, mercury and all the various different gases expand in very nearly simple numerical proportion. If scales of gas volume and mercury length are given uniform step, and made to coincide at 0° and 100°, the readings of the two corresponding thermometers nearly coincide when they are immersed in some vessel of water at any specific warmth-condition between 0° and 100°. Logically, something is thereby proved about the expansion of the gas with temperature, *when the length of a thread of mercury is arbitrarily taken as a measure of temperature, and when at the same time the volume of the gas held at constant pressure is arbitrarily rejected as an alternative mode of measurement of temperature.*

The accurate measurements of Dulong and Petit (1817)[15] showed that all gas thermometers give *nearly but not quite* the same value of the temperature of a given warmth-condition when the gas thermometers are standardised to read alike at two fixed points. This is true whether the thermometers are of the constant pressure or constant volume kind.[16] Experiment shows that the *nearly equal* records of the temperature of the given warmth-condition *tend towards identity* as the pressures of the gases used tend towards zero.

This is a practical reason for the preference of a gas thermometer, to give a standard temperature scale. But there is no Platonic kind of ideal temperature, above or beyond a given warmth-condition, to which our thermometers aspire and which they fail to register.

[15] W., pp. 33, 34. [16] W., p. 37.

It may here be repeated that we are always concerned only with a surely and accurately *reproducible*, generally *comparable* temperature scale; there is no question of a "true" or "natural" scale of temperature.[17]

The numbers which appear in the thermometric scales are misleading in that they suggest that temperature is a quantity, which it is not. Logically, the numbers are a plurality of fixed points, and they could be replaced by such. Nevertheless the numbers are useful, for they correctly indicate that the boiling-point of aldehyde (21°C) is warmer than melting ice (0°C) and cooler than boiling water (100°C). In short the numbers are an *ordinal* set.[18]

(g) *The definition of temperature*

After this analysis, Mach pauses to define the term temperature. Suppose the pressure of a gas is kept constant, at p_0, and that at different warmth-conditions the volume of the gas is

$$V_0,\ V_0(1+\alpha),\ V_0(1+2\alpha),\ V_0(1+3\alpha)\ .\ .\ .\ V_0(1+t\alpha)$$

then the corresponding temperatures are

$$0,\quad 1,\quad 2,\quad 3\quad .\ .\ .\ t,$$

there being a one-to-one correspondence between warmth-condition and temperature.

Equally well, T can be defined by an equation of the type

$$p = \frac{T}{k} p_0,$$

where the volume of the gas is kept constant at V_0. If k is 273, and if $p = p_0$ at 0°C, then on the T scale, 0°C corresponds to 273.

If in constructing the thermometer and defining temperature one is confined to expansion as an associated characteristic, even so there results a multiplicity of thermometers and scales which do not quite agree with each other, in the sense that the various instruments do not record precisely the same reading

[17] W., p. 51. [18] W., p. 67.

for a given single warmth-condition. Indeed the various temperature scales are arbitrary in a three-fold degree: there is the arbitrary selection of volume as an associated characteristic, the arbitrary selection of a specific thermometric substance such as mercury or hydrogen and the assignment of arbitrary numbers to two fixed points.[19]

(h) *A different associated characteristic*

In 1750 Richmann obtained the formula

$$t = \frac{m_1 t_1 + m_2 t_2}{m_1 + m_2}$$

for the temperature t of mixture of m_1 g of water at t_1 and m_2 g of water at t_2.[20] To Richmann himself the formula represented simply an experimental discovery concerning masses m_1, m_2 measured by the balance and temperatures t, t_1, t_2 recorded by a common liquid-in-glass thermometer. But the formula may be rewritten as a definition of temperature:

$$t \equiv \frac{m_1 t_1 + m_2 t_2}{m_1 + m_2}.$$

(Here and elsewhere in this book the symbol $\equiv$ means 'is defined to be'.) Then the temperature 1°C could be defined as that of a mixture of 99 g water at 0°C, and 1 g water at 100°C, the use of the usual fixed points being permissible:

$$1 \equiv \frac{(99 \times 0) + (1 \times 100)}{100}.$$

Each other point of the scale could be defined in similar ways. For convenience the temperature scale so produced is called here the Richmann scale.[21]

The Richmann scale depends arbitrarily on the properties of a specific substance, water. It is open to us to set up a second Richmann scale (Richmann scale II), which may have the same or equivalent fixed points, but for which, in the definition

$$t' \equiv \frac{m_1 t_1' + m_2 t_2'}{m_1 + m_2},$$

[19] W., pp. 43, 48. [20] W., p. 154. [21] W., pp. 183–5.

the quantities m_1 and m_2 are masses, not of water, but of alcohol. Mach goes so far as to draw a graph showing Richmann I temperatures plotted against Richmann II temperatures. Such a graph is never quite rectilinear.

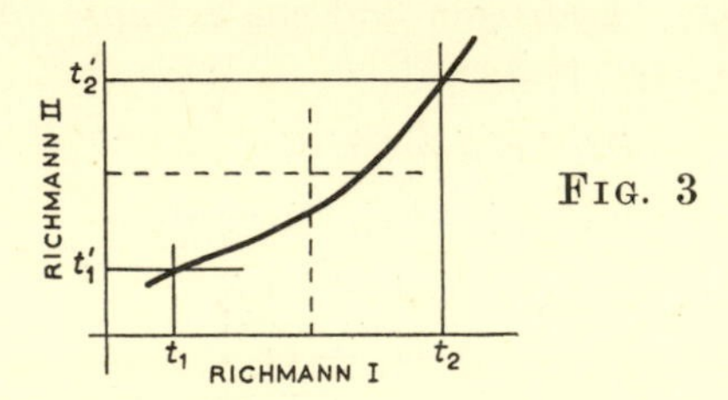

FIG. 3

Although the temperature t_1' corresponds to t_1, and t_2' to t_2, the temperatures

$$\frac{t_1 + t_2}{2} \quad \text{and} \quad \frac{t_1' + t_2'}{2}$$

will not correspond to a single warmth-condition. As Mach points out, the fact that experience teaches us that this graph is nearly rectilinear makes no difference to the logic of the discussion.

(i) *The idea of an absolute zero of temperature*

If we think again of the expressions for the volume of a gas at constant pressure and at successive temperatures 0, 1, 2, 3 . . . °C, i.e.

$$V_0, V_0(1 + \alpha), V_0(1 + 2\alpha), V_0(1 + 3\alpha) \ldots,$$

we may go on to inquire: At what temperature t will the volume of the gas become zero? t, so defined, is given by the equation

$$V_0(1 + \alpha t) = 0$$

and hence

$$t = -\frac{1}{\alpha},$$

which seems to be a kind of absolute zero of temperature. Equally for the constant volume scale, the absolute zero of temperature t could be defined by

$$p_0(1 + \alpha t) = 0.$$

Whatever plausibility such arguments may have, arises from the simple fact that, in classical physics, one cannot conceive of either a negative volume or a negative pressure.

Mach notes in the first place that this kind of game may be played with a different substance, and that, if so, a different answer is forthcoming. The gas thermometer indicates that the absolute zero is about −273°C; but a mercury thermometer would suggest that it is about −5000°C.[22] What kind of absolute quantity are we considering, if its value can be any arbitrary number?

That anyone conceived of an absolute zero at all is an accidental fact of the history of physics. Instead of the temperatures −2, −1, 0, +1, +2 on the Celsius constant pressure gas thermometer being represented by the volumes

$$V_0(1 - 2\alpha),\ V_0(1 - \alpha),\ V_0,\ V_0(1 + \alpha),\ V_0(1 + 2\alpha)$$

they could be represented by the volumes

$$V_0(1 + \alpha)^{-2},\ V_0(1 + \alpha)^{-1},\ V_0,\ V_0(1 + \alpha)^{1},\ V_0(1 + \alpha)^{2}.$$

A scale of this kind was in fact suggested by John Dalton. For a common value of α, the two scales would tally closely for a range of temperatures above and below 0°C.[23] But Dalton's scale would lead to no expectation of an absolute zero. If we write

$$V_0(1 + \alpha)^t = 0,$$

evidently $V_0(1 + \alpha)^t$ would tend to zero as t tends to $-\infty$. Using Dalton's thermometer we should have no more chance to catch up with the absolute zero, than had Zeno with the tortoise —and for the same reason.

It will be convenient to conclude Mach's theory of metrical temperature after his work on the first two laws of thermodynamics has been reviewed.

[22] W., p. 54. [23] W., p. 47.

4. Operational view of a concept

The late P. W. Bridgman defined the metrical concept simply as a "set of operations".[1] If, for example, Einstein's 'operations for length' differ from 'the common operations for length', then it should not tacitly be assumed that

$$\text{length}_{\text{common}} = \text{length}_{\text{Einstein}},$$

for strictly there are *two* concepts in the case.[2]

By the term 'operation' Bridgman means some kind of experimenting.[3] Since experiments can be broken down—at least roughly—into sense-perceptions, Bridgman's position is very similar to that of Mach:

> In origin the concept [of force] doubtless arises from the muscular sensations of resistance experienced from external bodies. This crude concept may at once be put on a quantitative basis by substituting a spring balance for our muscles. . . .
> The most fundamental of these [concepts of thermodynamics], which sets thermodynamics off apart from the simpler subjects, is probably that of temperature. In origin this concept was without question physiological, in much the same way as the mechanical concept of force was physiological.[4]

Bridgman's distinction between 'crude concept' and 'quantitative concept' corresponds to the distinction drawn here between 'shadow-concept' (pre-concept)' and 'metrical concept'. Mach's theory of the metrical concepts amounts, quite roughly, to the view that *some of them are sense-perceptions*; Bridgman's reference to a 'sensation' as actually being a concept at once shows him up as an ally of Mach.

5. Metrical 'non-O' concepts: Introductory

To Berkeley it is absurd to seek for something—substance, body, matter—lurking behind what appears as a table; the *being* of the table is no more than its *being perceived*. This part

[1] P. W. Bridgman, *The Logic of Modern Physics*, p. 5.
[2] Ibid., pp. 10, 24. [3] Ibid., p. 7. [4] Ibid., pp. 102, 117, 118.

of Berkeley's teaching is accepted both by Hume and Mach. So to Mach the notion of substance is merely a fiction, and 'a body' is no more than 'a group of sense-perceptions'.

Nevertheless, Mach is aware of the value of the 'substance concepts' as they are employed as powerful terms in the communication of physics. Mach is a good historian of his subject. For example, he understands how the notion of light as a substance inspired Kepler. Undoubtedly, Kepler considered light as a quantity of something "which diminishes in surface density when spread out on a larger surface and increases in density when collected, just as in the case of a cloud of dust, a drizzle or a coat of paint."[1] In 1604, by thinking of light in this way, Kepler was led to the view that the intensity of illumination of a surface by a bright point of light varies inversely as the square of the distance of the surface from the point. This is so easily accounted for by the idea of 'light substance' being spread uniformly over the surfaces of concentric spheres with the light point as their common centre—spread for example four times as thinly over the sphere of twice the radius—that it is difficult to disentangle the notions of intensity of illumination and quantity of light. But it is the former which properly is the inaugurating concept of optics.[2]

Mach's discussions of the metrical 'non-O substance concepts' —mass, quantity of heat, and energy—are masterly. Each is taken as 'conserved' substance, for the notions of 'substance' and 'conservation' are correlative. In the present chapter, one of these three concepts, the concept of 'quantity of heat', is considered in some detail. Mach was inspired by Kant's *Prolegomena* which is still the best of all philosophical interpretations of the substance conception. Kant's great monograph is also relevant to Mach's account of geometry as a kind of physics. Since it was a remark of David Hume which interrupted Kant's dogmatic slumber, it is proper to preface the section on Kant by one on Hume. Mach read the *Prolegomena* when he was a boy; the writings of Hume were more congenial to him, although he probably read no Hume before 1880.

[1] O., p. 17. [2] O., pp. 16–19.

6. The panorama of experience: Hume

When a moving billiard ball A collides with another ball B at rest, B begins to move. The momentum of A is shared with B and the total momentum of A and B together is the same as the original momentum of A. So the physicist is able to interpret the events in Newtonian terms. In his report, he is likely to affirm that the motion of A has *caused* the motion of B. Before a man has seen this experiment or anything like it, the outcome could be any conceivable event or set of events. For example when A touches B, A could stop dead, evaporate in a green flash, or rise into the air.[1] Thus before one has the experience and other such experiences, there is no question of reasoning about the events. One must have sense-perceptions before one can think. 'Feeling', as Hume calls it, must precede thinking.

> . . . it is impossible for us to *think* of any thing, which we have not antecedently *felt*, either by our external or internal senses.[2]

Hume is determined to report what happens in this case, *and only what happens*. If A *causes* the motion of B, this causation is not seen. Whatever they may be, *causation* and *necessary connexion* are not phenomena. What is enacted before the observer is a *panorama*, first this and then that, "entirely loose and separate".

> All events seem entirely loose and separate. One event follows another; but we never can observe any tie between them. They seem *conjoined*, but never *connected*.[3]

Again, and even more clearly:

> The impulse of one billiard-ball is attended with motion in the second. This is the whole that appears to be *outward* senses. The mind feels no sentiment or *inward* impression from this succession of objects: Consequently, there is not, in any single, particular instance of cause and effect, any thing which can suggest the idea of power or necessary connexion.[4]

Why then do we nevertheless at last begin to feel that the motion of B has been caused by A, and that there is a connexion between the successive events? This is brought about—such is Hume's answer—by a long "course of uniform experience".[5]

[1] D. Hume, *Enquiries*, p. 48. [2] Ibid., p. 62. [3] Ibid., p. 74.
[4] Ibid., p. 63. [5] Ibid., p. 79.

Billiard balls are seen at different times in different places by different people; other kinds of ball also seem to behave in the same kind of way.

The first time a man saw the communication of motion by impulse, as by the shock of two billiard balls, he could not pronounce that the one event was *connected*: but only that it was *conjoined* with the other. After he has observed several instances of this nature, he then pronounces them to be *connected*. What alteration has happened to give rise to this new idea of *connexion*? Nothing but that he now *feels* these events to be *connected* in his imagination, and can readily foretell the existence of one from the appearance of the other.[6]

The man, so conditioned by his experience, is prepared to call one event *cause* (motion of A alone) and the other event *effect* (motions of A and B). Further he is prepared to prophesy future events of the same kind.[7] The multiplication of instances then has no logical force. Experience requires the multiplication of instances merely to establish a custom or habit of the mind.

All inferences from experience, therefore, are effects of custom, not of reasoning.[8]

So Hume is led to the classical statement of the problem of induction. A rational process cannot require "a hundred instances". Learning from experience cannot then be a "process of reasoning".

Nothing so like as eggs; yet no one, on account of this appearing similarity, expects the same taste and relish in all of them. It is only after a long course of uniform experiments in any kind, that we attain a firm reliance and security with regard to a particular event. Now where is that process of reasoning which, from one instance, draws a conclusion, so different from that which it infers from a hundred instances that are nowise different from that single one? . . . I cannot imagine any such reasoning.[9]

Hume's argument is stimulating. If we came across a geometer who first proved a theorem with a small pink triangle, then a large green one, then on both Tuesday and Wednesday, then in Scarborough as well as in Hull, and so on, we might send him back to the mad-house, or, more respectfully, remind him

[6] Ibid., p. 75. [7] Ibid., p. 74. [8] Ibid., p. 43. [9] Ibid., p. 36.

that his subject is mathematics and not physics. *Prima facie*, it seems clear that the activity called physics is no less mad than that of this unusual geometer. Hume's principle, that the panorama of experience is not in itself rational and that it is necessarily prior to the reasoning processes of a science such as physics, is inexorable.

Mach accepts this principle, although his exposition of it is briefer and less clear than that of Hume:

> There is no cause nor effect in nature; nature has but an individual existence; nature simply *is*. Recurrences of like cases in which A is always connected with B, that is, like results under like circumstances, that is again, the essence of the connection of cause and effect, exist but in the abstraction which we perform for the purpose of mentally reproducing the facts. . . .
>
> Hume first propounded the question: How can a thing A act on another thing B? Hume, in fact, rejects causality and recognises only a wonted succession in time. . . . The notion of the *necessity* of the causal connection is probably created by our *voluntary* movements in the world and by the changes which these indirectly produce, as Hume supposed. . . .[10]

7. Kant's Prolegomena: Paraphrase and commentary

Kant draws the important distinction between analytic and synthetic judgements. Analytic judgements do not enlarge knowledge, they unfold or develop what is already known. They are logically certain, but contain no genuine discovery. They can be called explicative.

Synthetic judgements, on the other hand, enlarge knowledge or are ampliative.[1] A science such as physics is based on experience. A judgement of experience must always be synthetic, for if there were no enlargement of knowledge there would be no experience. As Kant says, "experience itself is nothing other than a continual joining together (synthesis) of perceptions".[2] Mach himself might well have written these words.

The experiential knowledge known as physics is what Kant

[10] M., pp. 580–1.
[1] I. Kant, *Prolegomena*, translated P. G. Lucas, Manchester: 1953, p. 16.
[2] Ibid., p. 30.

calls 'synthetic *a posteriori*'. But a most important and original part of Kant's teaching is that there can also be synthetic judgements which have *a priori* certainty, and which "have their origin in pure understanding and reason."[3] If this is so, *some* synthetic judgements can be 'certain'. Kant maintains that *all* mathematical judgements are of this kind, that they are *a priori*, synthetic, and certain.[4]

Emile Meyerson gives an illustration of the synthetic nature of geometrical reasoning and for this purpose cites Euclid's proof of the theorem of Pythagoras.

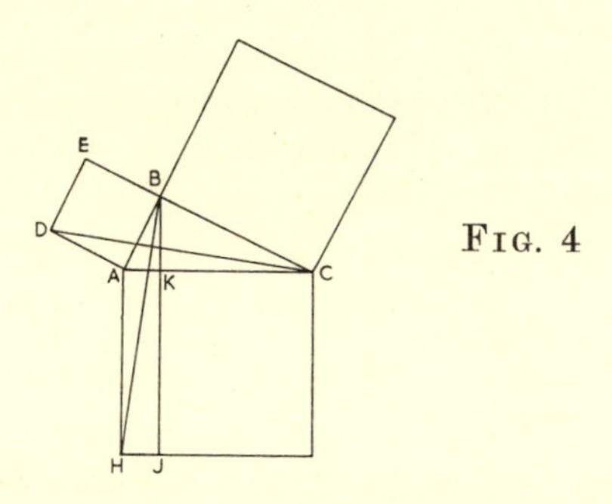

Fig. 4

The right-angled triangle ABC is depicted in Fig. 4, AC being the hypotenuse. As Meyerson indicates[5] the two triangles ABH, DAC are equal to each other in a sense different from that in which the square ADEB and the rectangle AKJH are equal. The square and rectangle are mediately judged to be equal because both areas are twice the area of the triangle DAC or ABH. The triangles are 'more identical' than the rectangles because they can be exactly superimposed. In the course of the proof, *there is the decision to accept equality of two areas which are not equal in the sense of being exactly superimposable.* This is, I think, an unanswerable proof that mathematics is not entirely analytic, and that synthetic judgements are made in mathematics. Meyerson develops the argument by an example from algebra.[6]

[3] Ibid., p. 17. [4] Ibid., p. 18.
[5] E. Meyerson, *De l'Explication dans les Sciences*, Paris: 1927, pp. 146–7.
[6] Ibid., p. 139.

Kant's teaching that mathematical judgements are synthetic *a priori* can only be understood by reference to his further teaching concerning space and time. In Kant's language the term 'intuition' is not a synonym for the term 'sense-perception.' Intuitions constitute a genus, within which sense-perceptions constitute a species. Intuitions may be impure (empirical) or they may be pure (*a priori*). The former kind are sense-perceptions; examples of the latter kind are the pure intuitions of space and time. Although, as A. D. Lindsay said,[7] Kant puts "space and time on the side of the *given*", yet he requires us to accept that space is an *a priori* pure intuition. Space and time are "forms of our sensibility". Because they are 'forms' they can be *a priori*; in so far as they are 'of our sensibility' they are 'given'. This is difficult doctrine; a few words from Kant may help the reader:

. . .; for if everything empirical, namely what belongs to sensation, is taken away from the empirical intuitions of bodies and their changes (motion), space and time are still left. These are therefore pure intuitions, which are the ground *a priori* of the empirical intuitions, and hence can never be taken away themselves, but prove, precisely by being pure intuitions *a priori*, that they are mere forms of our sensibility which must precede all empirical intuition, i.e. perception of real objects, and in conformity with which objects can be known *a priori*, though indeed only as they appear to us.[8]

Moreover, according to Kant the two main kinds of mathematics, geometry and arithmetic, are based respectively on the pure intuitions of space and time. The three vertices of a triangle are, as it were, contemporary; but the number 2 is counted out *after* the number 1. So:

Geometry is grounded on the pure intuition of space. Arithmetic forms its own concepts of numbers by successive addition of units in time: . . .

Pure mathematics, as synthetic knowledge *a priori*, is only possible because it bears on none other than mere objects of the senses, the empirical intuition of which is grounded *a priori* in a pure intuition (of space and time), . . .

That complete space (which is not itself the boundary of another

[7] A. D. Lindsay, *Kant*, Oxford: 1934, p. 67.

[8] I. Kant, *Prolegomena*, p. 39.

space) has three dimensions, and that space in general cannot have more, is built on the proposition that not more than three lines can intersect at right angles in a point. This proposition cannot be shown from concepts, but rests immediately . . . on pure intuition *a priori*.[9]

Mach admires Kant's view that *we* necessarily provide inescapable forms or modes of perception when we observe the external world:

According to the doctrine of another great philosopher, Kant, time and space lie not so much in things, as *in us*; they are inescapable modes of perception in which we necessarily observe the external world and the processes within ourselves.[10]

Mach believes that geometry is a species of physics. He therefore welcomes Kant's doctrine that space is *given*.[11] Unwisely, in my view, he discards Kant's distinction between pure and empirical intuitions. So to Mach, a space-sensation (Raumempfindung) is just one amongst other sense-perceptions: "Colours, sounds, temperatures, pressures, spaces, times, and so forth, are connected with one another. . . ."[12] Yet he is compelled to admit that the elements of space and time have a distinguishable difference:

A body is perceived in a certain *place* and at a certain *time*. Consequently a space *sense-perception* and a time *sense-perception* belong to the sensuous complex which represents the body. The fact that a body is *capable of being moved about* indicates *a certain lack of stability* of the space and time elements within the complex, and that the other elements are relatively *more stable* units of the body.[13]

I take an orange with me from Hull to Bridlington. The taste, smell and orange colour stick tightly together but the 'space of Hull' breaks away from the orange and is replaced by some 'space of Bridlington'; the 'time 3.30 p.m.' also breaks away and is replaced by '4.15 p.m.' The molecule, 'an orange at Hull at 3.30 p.m.', has weak valency bonds linking the atoms of space and time, but a strong valency bond between the colour and the smell atoms. The last quotation shows Mach at his

[9] Ibid., pp. 39–41. [10] PP., p. 492. [11] S., p. 34. [12] A., p. 2.
[13] W., p. 424.

worst. The opinions violate common-sense. We do not see both the orange and the 'nothingness' in which it is extended, in one and the same way. Even W. Ostwald, whose views resembled Mach's in many respects, would have objected that there must be, in an act of genuine sense-perception, a flow of energy from the object to be received by the sense-organ. In the intuition of space *by itself* there is no such flow. We need not decide here whether Kant is entirely correct in describing space and time as pure *a priori* intuitions, but evidently the experience of space and the experience of redness are qualitatively different, requiring two different designations. Kant is certainly correct in finding that *space and time are the ground of all sensuous experience and not just a part of it.*

The *Verstandesbegriffe* of Kant, the pure concepts of the understanding, are very general principles in accordance with which the human mind can think. They are so general that no metrical concept of physics can be identified as one of them, nor can the definition of any single metrical concept of physics be deduced from them. The metrical concepts and laws of physics are sought out to satisfy and exemplify these very general concepts of the understanding. In this way, the metrical concepts and laws of physics are *prescribed by the mind to nature.* As the French philosopher Boutroux said, the Kantian pure concepts of the understanding are not laws of nature but *moulds* into which these laws must be *poured.*[14]

In the *Critique of Pure Reason* Kant lists twelve of these pure concepts of the understanding, or categories, as they are also called. They are divided into three groups of four. The categories of

'substance'

and 'cause and effect'

are two closely related ones taken from Kant's third group. According to these categories,

Whatever changes occur, something must persist unchanged:

and *When a change occurs, the effect follows the cause.*

[14] E. Meyerson, *Identity and Reality*, translated Kate Loewenberg, London: 1930, p. 147.

One cannot *deduce* from Kant's principles the law of conservation of energy, but one can—guided by those same principles—*require* some such kind of conservation. *Experience* alone cannot establish that, when a book of mass m is lifted through a height h in the earth's gravitational field characterised by the acceleration g, the system has gained potential energy mgh, but the *human mind* requires this to be the case in order that its experience may become intelligible. Or again, in the example of Hume's billiard balls, it is impossible to *see* momentum or to experience its conservation. Momentum is, as P. W. Bridgman said, no more than "a set of operations" based on a clever idea of Newton. It satisfies our minds, because it enables the empirical billiard ball law to be poured into the mould of a Kantian *a priori* category. Examples may be drawn from sciences other that physics. In the chemical change represented by 'the equation'

$$2Hg + O_2 = 2HgO,$$

a silvery metal and colourless gas give way to a red crystalline solid; but the mass of the red solid equals the combined masses of the silvery metal and the colourless gas. The change is *intelligible*, because it is—in part—*denied*. The very term 'equation' as used by chemists is Kantian; the very name 'mercury oxide', which suggests that somehow the mercury and oxygen are still there in the red solid, is Kantian too.

Even a slight understanding of Kant's thought is most enlightening. One wonders why one could ever have been so stupid as to imagine that one might *see* necessary connection. Necessary connection must lie in the mind. Kant's theory enables us to find necessity in physical as well as mathematical thought.

We are none the less really in possession of pure natural science, in which laws to which nature is subject are propounded *a priori* and with all that necessity which is required for apodictical propositions. . . . But there are several among the principles of this general physics which really have the universality that we demand, such as the proposition: *that substance remains* and is permanent, that *everything that happens is* always previously *determined* according to constant laws *by a cause*, etc. These really are universal laws of nature which subsist wholly *a priori*.[15]

'Necessity' characterises both mathematics and physical science; in the former, it arises from the *a priori* intuitions of space and time, and in the latter from the *a priori* concepts of the understanding. At this deep level, Kant broadly identifies mathematics and physics.

Mach is obliged to reject Kant's doctrine of the categories. If there is a single web of elements, if the ego is simply a local concentration in this web, if the ego has the same status as the kitchen table, and if the ego is or includes the mind, there is then no basis for a theory which distinguishes between a sense-perception and a pure *a priori* intellectual concept. It does not occur to Mach that it may be wiser not to call two things by the same name until it is established that they are the same.

Kant uses the term 'experience'[16] in a technical sense. The panorama of impressions (Hume) is not experience. The impure intuitions have to be 'thought'[17] before they become experience. There is therefore a Kantian equation of this kind:

$$\text{Intuition} + \text{Thought} = \text{Experience.}$$

The term 'thought' refers to the *a priori* concepts of the understanding which, *on being added to sense-perceptions transform them into genuine experience*. Experience is thus a philosophical compound of two qualitatively different elements. I quote one sentence from the *Prolegomena:* ". . . experience [is] . . . the product of the understanding out of the materials of sensibility."[18] Even more clearly in the *Critique:*

Thoughts without content are empty, intuitions without concepts are blind. . . . These two powers or capacities cannot exchange their functions. The understanding can intuit nothing, the senses can think nothing. Only through their union can knowledge arise. But that is no reason for confounding the contribution of either with that of the other; rather is it a strong reason for carefully separating and distinguishing the one from the other.[19]

[15] I. Kant, *Prolegomena*, p. 53. [16] Ibid., p. 56. [17] Ibid., p. 64.

[18] Ibid., p. 77.

[19] I. Kant, *Immanuel Kant's Critique of Pure Reason*, translated Norman Kemp Smith, London: 1958, p. 93.

Kant maintains further that when sensible intuition is turned into experience by the addition of the pure *a priori* concepts of the understanding, the result is a judgement of experience at once necessarily and objectively valid.[20]

Unlike Berkeley and Mach, Kant believes that as the ground of all our experience, there are things-in-themselves. They correspond roughly to the Cartesian primary qualities, but differ from these in that they are quite outside any possible intuition. They are also outside the possibility of our thinking about them by the use of the categories. They serve however a very important philosophical purpose: they save Kant's doctrine from being a *pure idealism*. There are things other than minds or Mind.

. . . our sensible representation is in no way a representation of things in themselves, but only of the way they appear to us . . . things are given to us as objects of our senses situated outside us, but of what they may be in themselves we know nothing; we only know their appearance, i.e. the representations that they effect in us when they affect our senses. . . . Can this be called idealism? It is the very opposite of it.[21]

Mach inquires in the *Analysis of Sensations* whether it is possible to take from a common object, a lump of sugar for example, its properties, e.g.

whiteness
sweetness
roughness
weight and so on,

one by one, and then whether there would remain a lump of sugar without any of these properties.

. . . it is imagined that it is possible to subtract *all* the parts and to have something still remaining. Thus naturally arises the philosophical notion, at first impressive, but subsequently recognised as monstrous, of a "thing-in-itself", different from its "appearance", and unknowable.[22]

Naturally from Mach's point of departure, that the lump of sugar is nothing more than a group of sense-perceptions, when

[20] I. Kant, *Prolegomena*, p. 57. [21] Ibid., pp. 43, 45. [22] A., p. 6.

these sense-perceptions are all subtracted, there can be nothing left. There is no question and there is no answer. Kant's *Ding an sich* does not arise in the way Mach suggests. It is a metaphysical object. What it is in itself we cannot know, because our knowledge must be a compound of intuition and understanding. The *Ding an sich* can neither be intuited nor thought.

8. The four aphorisms

Mach considers it more appropriate to replace Kant's two famous aphorisms, that

Thoughts without content are empty;
and Intuitions without concepts are blind:

by two others. I quote from *Knowledge and Error*:

Kant says: "Thoughts without contents are empty, intuitions without concepts are blind." Possibly we might more appropriately say: Concepts without intuitions are blind, intuitions without concepts are lame.[1]

Let us consider the four aphorisms of Kant and Mach, taking those of Kant first. The categories of the understanding need sense-perceptual material to work upon. The empty thoughts are the moulds into which experience is to be poured. That sense-perceptions without concepts are blind is Kant's recognition of the truth of Hume's original criticism. Sense-perceptions without an understanding mind to interpret them are a meaningless succession or panorama. The eye seeing, sees but does not understand. It is a paradox, the eye which *sees only* is quite blind.

When Mach affirms that concepts without visual perceptions are blind, he means much less than Kant. The concepts in question are the concepts of physics, e.g. mass and energy; they are not the pure *Verstandesbegriffe* of Kant. An ordinary physical metrical concept created freely without reference to experience, the most important part of which is visual perception, would be blind in that it would not help the physicist to

[1] E., p. 385.

find his way about in the realm of physical science. This is important, worth saying and it is a part of Mach's *credo*. The metrical concepts of physical science are created out of experience, or they are created in order to interpret experience. When however Mach invites us to compare

Concepts without intuitions are blind (Mach)
with Thoughts without content are empty (Kant),

he is perpetrating a philosophical pun. The *Massbegriff* of physics is not the same kind of idea as the *Verstandesbegriff* of Kant's metaphysics.

The second of Mach's sayings, which could also be correctly translated as 'Visual perceptions without concepts are lame', is his admission that the panorama of experience, taken as the whole of physical science, would add up to a limping or stumbling physics. Although Mach himself has written well and at length on the theoretical or intellectual element in science, he is unwilling to go all the way with Hume and Kant. *Sense-perceptions alone, according to Kant, could not constitute any kind of physics at all.* He is right. Had Kant meant merely that sense-perceptions without metrical concepts do not add up to any kind of physics, he would still have been right. What he does mean is more profound, and leads his readers into a systematic metaphysics of nature where Mach feels himself unable to follow. All four statements may therefore be defended as correct—I personally accept them all—but the coverage of Mach's two statements is comparatively small.

9. Mach and the notion of substance

Mach maintains, correctly in my view, that all the legitimate concepts of physics are related to experience. He also recognises that there are at least two kinds of relationship between concepts and experience. Length, force, temperature, intensity of illumination are more directly related to experience; mass, quantity of heat, energy-comprehensive, mechanical energy, quantity of electricity are related much less directly. It is not quite correct, but it is not meaningless to say that 'I feel the temperature of the fire'; it is quite wrong to say that 'I feel the heat of the fire'. It is not quite correct to say that 'I felt the heavy weight'; it is entirely wrong to say 'A feels more massive than B' or that 'A looks more massive than B'. In the same kind of way, a book on a high shelf does not 'look' more energetic than when it is on a low shelf, but it is said to have more potential energy in the former case. All the metrical concepts are related to experience, and are also intellectual constructions. The second of the two characteristics is however more dominant in respect of the 'non-O' concepts. The definition of the 'non-O' concepts is a work of art, the greatest creative act of the mind of the scientist.

It is characteristic of some 'non-O' concepts that, in a limited context at least, they stand for persistent quantities of a substance, regarded, again perhaps in a limited context, as indestructible. They are like matter or stuff; they are stuff-concepts. Black's own term for 'quantity of heat' was 'matter of heat'. Mach's attitude to these concepts is very similar to Berkeley's attitude towards matter. Berkeley considered that there is no existing matter 'behind' what is perceived by the minds of men and the mind of God; in the same way Mach considers that, although the 'non-O' concepts and especially the stuff-concepts, are of great economic value in the direct or indirect description of nature, they do not represent any actual existing stores of substance. This is so true that, to some extent at least, what we choose to regard as 'an indestructible fluid' is a matter of taste. My designation of the concepts quantity of heat and mass as 'non-O' is a fairly accurate summary of Mach's teaching. Between our minds and such concepts there is no direct link of sense-perception; in Mach's philosophy we can find neither mass nor heat in the one web of the elements.

How, so Mach inquires, does the notion of substance arise in the first place? Primitive man finds that certain sense-perceptions and experiences 'go together' as it were; he learns economically to ask for *water* rather than to present a list:

wet
colourless
thirst-quenching   and so on.

There is no doubt that in this quite simple psychological sense the idea of the *substance water* has great economic value.[1] The idea of 'substance' and the idea of 'persistence' are correlative. The correct mathematical description of "a cube with trimmed corners"[2] would be formidably difficult. Such a figure is not a cube at all; in order to economise effort, to run the business of ordinary life economically, we decline to admit that there is a cube no longer, we think of the cube as *still there*—although modified.

Even such a simple example as this differs in degree but not in kind from the principle of conservation of energy. The human mind has an "instinctive craving for concepts involving quantities of things".[3]

> All our efforts to mirror the world in thought would be futile if we found nothing permanent in the varied changes of things. It is this that compels us to form the notion of substance, the source of which is not different from that of the modern ideas relative to the conservation of energy. . . . Where does heat come from? Where does heat go to? Such childish questions in the mouths of mature men shape the character of a century.[4]

All the main branches of physics afford numerous examples of the use of substance concepts. Mach's brilliant analysis of Coulomb's interpretation of the division of force in an electrostatics experiment as the flowing over of the electrical fluid-substance from one body to another, to which there is a just analogy in the imagined flow of heat fluid-substance from hot water to cold water on mixing, has been sketched out in the previous chapter.

Mach will not permit us to think that oxygen is 'more of a substance' than latent heat; he prefers to draw a clever analogy between them:

> Heat is a substance just as much as oxygen is, and it is not a substance just as little as oxygen. Substance is possible phenomenon, a convenient word for a gap in our thoughts. . . .
>
> If we explode a mixture of oxygen and hydrogen in a eudiometer-tube, the phenomena of oxygen and hydrogen vanish and are replaced by those of water. We say now that water *consists* of oxygen

[1] PP., p. 471 (untranslated paper, 1910). [2] P., p. 201. [3] M., p. 329.
[4] P., pp. 199–200.

and hydrogen; but this oxygen and this hydrogen are merely two thoughts or names which, at the sight of water, we keep ready, to describe phenomena which are not present, but which will appear again whenever, as we say, we decompose water. It is just the same case with oxygen as with latent heat. . . . If latent heat is not a substance, oxygen need not be one.[5]

On the other hand, one might add, if heat *is* a substance, then may oxygen be one also. The example, brilliant in itself, is parallel to Hume's discussion of the colliding billiard balls. In Hume's example momentum is thought of as a substance passing from one ball to the other, also as being conserved. How soon the exercise of scientific interpretation goes far beyond what is simply observed is even better shown by Mach's chemical example than by the classical illustration of Hume.

Mach's historical insight leads him to give a full and candid account of the substance-concepts, although he is perhaps unnecessarily intolerant of substance in the form of atoms and molecules. In the end, Mach believes that we can and that we ought to learn to do without substances, bodies, atoms and matter. We ought to learn to purify our minds from these mere *adjuncts*, we ought to return to our primary task of examining the connections between the elements given to us in the great web of sense-perceptions:

> It is basically thus a *constancy of connection* between sensuous elements, with which we have to do. If one could *measure* the entire set of sensuous elements, one could then say that the body *consists* in the satisfaction of certain equations which obtain between the [measured] sensuous elements. . . .
> In so far as the *conditions* of a phenomenon become recognised, the impression of "a stuff" recedes into the background.[6]

It is interesting to notice that Mach thinks measurement is one of the means which facilitate our retreat from the myth of substance. What we must avoid at all costs is the idea of substance as 'reality behind appearances'. Mach has no time for the *Ding an sich* of Kant, or for the atoms of Democritus which exist in reality.[7]

[5] G., p. 48. [6] W., pp. 423, 429.
[7] K. Freeman, *Ancilla to the Pre-Socratic Philosophers*, Oxford: 1952, p. 93.

10. The fable of the man and the column

A man walks behind a stone column[1] and reappears again on the other side. We see the man on one side of the column, later we see half his body and the column, then the other half of his body and the column, and finally, separately again.
In such an example as this it would be intolerably laborious to describe what we actually see; it is certainly a great economic advantage to be able to declare simply that the man has walked behind the column. In doing so we evidently think of the man as a 'persistent substance' or as a 'constant quantity'.

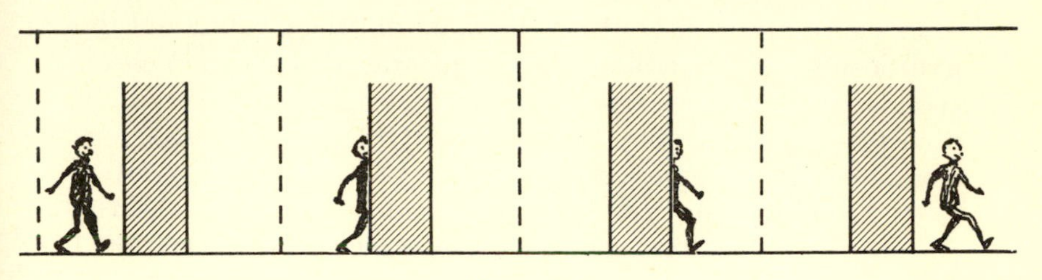

FIG. 5

The comparison between the man walking behind the column and the general fact of chemical change is instructive. When mercury and oxygen combine chemically they disappear from our view. We call the red crystals which appear mercuric oxide as a kind of act of faith. Mach would say we do it to ease our minds. When the mercury and oxygen reappear 'from behind the column', when they re-emerge from the strange dimension of chemical change as it were, we feel our faith has been justified and our minds further reassured.

Nevertheless Mach's interpretation of the man walking behind the column is not consistent with his philosophy of physics as a whole. In the first chapter of this study I have noted that, for the effective enterprise called physics, a great multitude of observers is needed, and that even their combined work needs the supplementation of the 'strolling sense-perception'. Suppose then, as Mach should have allowed, we call upon two observers, one in front of the column and one behind it. Then as the man

[1] P., p. 202.

walks behind the column, one of the observers keeps him in view all the time. It is a *more adequate* account of the *combined* experience to say that the man walks behind the column, than it would be to say that the first observer alone first sees a man on the left, then he sees a column and no man, then an emergent nose on the right edge of the column, and so on. Without any explicit reference to Kant, the former description is prfeerred to the latter; *more* of what has been experienced by *more observers* is *described.* It is somewhat bizarre to assert, as Mach does, that the former description is preferred, only because it is the more economical.

In the comparison between the man and the column and the hydrogen/oxygen reaction, the disappearance of the man corresponds to the disappearance of the hydrogen and oxygen. In their place however water is formed, and it has positive sensuous characteristics—wetness and the like—of its own. What interested Lavoisier however in such chemical reactions was the persistence of mass *throughout* the whole set of transformations:

$$\left.\begin{matrix}\text{hydrogen and}\\ \\ \text{oxygen}\end{matrix}\right\} \rightarrow \text{water} \rightarrow \left\{\begin{matrix}\text{hydrogen}\\ \\ \text{and oxygen.}\end{matrix}\right.$$

The mass of the original two gases, the mass of the water and the combined mass of the regenerated elements are all equal. The true analogy therefore is between the man who is 'there' all the time even when he is behind the column, and a mass M which remains constant and which may represent the mass of the two gases, or of water, or of portions of each. As I have pointed out in the early part of this chapter, mass is absolutely imperceptible and has no *direct* relationship to experience or sense-perception. What strictly persists is therefore invisible, and the analogy with the man out of sight is a good one. But it is not good for the reasons which Mach suggests. It is good because during the chemical change, the 'non-O' quantity mass remains constant. Kant said that only the permanent can change, and in the end Mach's fable is a simple Kantian exercise. When the man emerges again from behind the column he must be in some sense the same man as he was before. John Smith cannot go from Hull to Scarborough for his holidays unless he is, in some sense, still John Smith when he gets out of the train at Scarborough. In the same way water is called hydrogen oxide, not only because hydrogen and oxygen can be regenerated from it, but also because the mass of water is equal to the mass of the two gases. The primary philosophical significance of the sign of equality in the chemical *equation*,

$$2H_2 + O_2 = 2H_2O,$$

is Kantian.

11. Mach and Black

Mach gives an admirable account of the work of Black on quantity of heat and various examples of latent heat, with generous references also to his predecessors and contemporaries.[1]

Black's conception of heat as a fluid standing at the same level or temperature in all the common non-living objects in a room, like the water in Pascal's vases, has already been mentioned. Very correctly, Black refers to the equilibrium of heat. He distinguishes quite clearly between the strength, intensity, level of heat or temperature and the quantity of heat. The latter indeed is proportional to the mass of the substance,[2] and the former is independent of the mass of the substance. This fact is emphasised by Black's term 'the matter of heat', a term which in Mach's view is no more objectionable than the analogous terms 'matter' or 'atom'. When two different substances at different temperatures, or two samples of the same substance at different temperatures, are mixed together, the equilibrium of heat is restored by the flow of a quantity of heat from the warmer to the cooler substance. Black assumes that the heat lost by the one must be exactly equal to the heat gained by the other, this simply being an aspect of any good substance hypothesis. The philosophical status of the proposition that

$$\text{heat lost} = \text{heat gained},$$

[1] W., pp. 153–81. [2] W., p. 156.

is clarified by Mach's account.[3] We have no sense-perception of heat at all; it is not an item in our experience, and it does not appear in the great web of the elements which make up the 'One' which is the 'All'. It is, in Mach's language, a mere *adjunct* (Zutat) concept; or, in my language, it is a 'non-O' concept. Experience can neither enforce us to reject nor to accept the proposition that

$$\text{heat lost} \equiv \text{heat gained.}$$

We accept it, because *we like our substances to be conserved*. It is, as Humpty Dumpty said, a question as to *who* is master, the answer to which question is that *we* are. The simplest case of the method of mixtures, where two lots of water at different temperatures are mixed together, is illuminated by Black's view of temperature as a level, and by one of the most beautiful analogies in classical physics. As Richmann found, the temperature t of the mixture is given by

$$t = \frac{m_1 t_1 + m_2 t_2}{m_1 + m_2},$$

where the symbols have the same meanings as before. With this is compared the formula for the height h of the centre of mass of two masses m_1 and m_2 at heights h_1 and h_2 above the base line OO (Fig. 6):

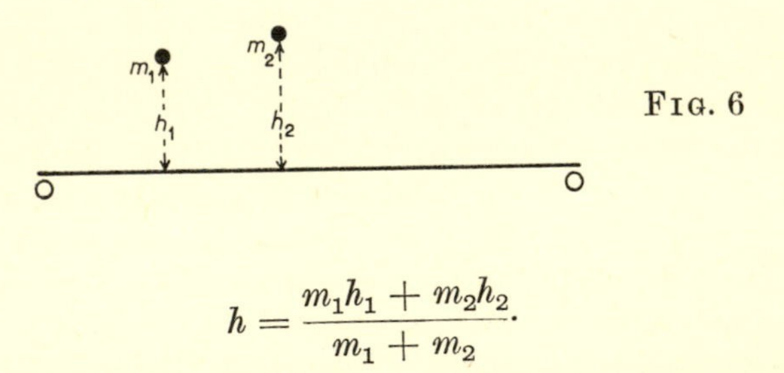

Fig. 6

$$h = \frac{m_1 h_1 + m_2 h_2}{m_1 + m_2}.$$

The two expressions have exactly the same mathematical form.

[3] The reader is reminded that $\equiv$ means 'is defined, or conventionally made, to be equal'.

Black discovered the sheer specificity of heat some time before 1765. His experiment on mixing equal masses of hot gold and cold water is well designed, and the specific heat ratio of the two substances is defined simply as the inverse ratio between the two temperature changes. Gold has such a small capacity for heat (Fig. 7) that its loss of a quantity Q causes a marked diminution of level (water poured out of narrow vessel A); water has such a great capacity for heat that, although it receives all of Q, the level rises very little (water received by vessel B).

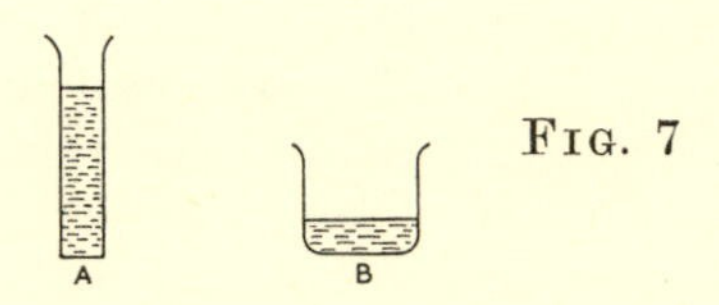

FIG. 7

The concept of specific heat is secondary, as it were, to that of the substance of heat; it is artificial and designed to 'save' the notion of the conservation of heat substance.

When water boils a great quantity of heat is pumped into the water. But instead of the temperature rising an event of an entirely different kind, viz. a change of state, occurs. The conception of heat hidden in the steam, hidden from us—insensible heat—and also hidden from our thermometers, is very bold. It is congenial to our minds, because by it the notion of the conservation of heat is 'saved'. Moreover the *latent* heat can become *sensible* again, for the condensation of steam to water at the same temperature will give out heat capable of warming something. Mach's vivid image of the man reappearing from behind the column is serviceable again.

The constancy of the quantity of heat was precisely a concept which had become agreeable to the mind, and which—had it been entertained metaphorically and not literally—would have been no hindrance to later investigations, as in fact it did turn out to be.

But with the thought that *processes of cooling down* might not necessarily be compensated by processes of warming up, but could be *compensated* by physical events of an *entirely different character*, Black made a significant step in that direction of thought which characterises the *thermodynamics* of our own time [1896], that discipline which recognises a connection between thermal events and [arbitrarily selected] physical events of any kind at all.[4]

This is scientific criticism of the highest order. If a physical change from liquid to gas can appear as a quantity of heat in a constant sum of heat terms, so, a century later, may men like Kelvin and Clausius regard work done by a heat engine as a heat term. In the Carnot reversible heat engine the heat given up to the sink is less than the heat taken from the source. The differential is proportional to the mechanical work done by the engine. Why should not Black's bold thought be applied here too? Why should the mechanical work not be measured in heat units, and why should it not be taken as a new species of latent heat? Change of state and mechanical work are alike unrevealed to our thermometers and to our sense-perceptions; they have an equal right to the description 'latent heat'. They are, as Mach says, of a character entirely different from a rise in temperature. The historical fact that it was Black who took the first vital step in modern thermodynamics has rarely been acknowledged. Mach's elucidation of the history of heat and thermodynamics is only a little less important that his criticism of Newton's dynamics.

12. Heat as a conserved substance

The unit of heat[1] may be defined as that amount of heat which warms 1 kg of water through 1°C, or—more precisely—which warms it from 0°C to 1°C. It does seem rather obvious that if there are n kg of water, then n times as much heat will be needed. It is not obvious in the same way that by the cooling of 10 kg of water 1°C, 10 kg of some other water can be warmed by 1°C; or that 5 kg of some other water can be warmed by 2°C and so on. These are not logical or necessary truths; they are experiential or experimental facts, and they are not quite correct. From such facts, the idea that the quantity of heat taken from or communicated to water can be measured by the product

$$\text{mass of water} \times \text{change in temperature}$$

[4] W., p. 194.

[1] W., p. 182.

is derived. The uncritical learning and teaching of this physics, regarded wrongly as both dull and easy, has brought logical discomfort to many good critical minds.

But why should anyone call the product

mass of water $\times$ change in temperature

a *quantity*?[2] Referring back to the first illustration, if we do something to each of n kg of water in turn, it seems natural to assume that the n kg contain 'n times as much of something' as 1 kg does. This is only partially satisfactory. Mach asks us to imagine six equal cylinders, with their axes vertical, standing before us on the table. One cylinder is rotated about its vertical axis 10° in the sense of the pointers of a clock. Then, in turn, the same is done to the other five. When the operations are complete, the cylinders look the same as they did before.

> Shall I call what the cylinders have now been given a *quantity*? Shall I say that the six cylinders contain six times as much of the quantity as one cylinder does?[3]

Evidently both these questions are to be answered in the negative, yet it is undeniable that one has 'done something' six times to six equal things. It is well worth while to compare the mixing of hot and cold water operation with the operation of turning the cylinders. When a man turns the cylinder round, it cannot be denied that he has conveyed a quantity of turning to the cylinder, although this sounds odd and artificial. The reason is simply that the man is not left with a kind of deficit of turning on his hands, as it were. In the heat process, quite often, although not always, as one body gets hotter another gets cooler. It is a very simple observation in everyday life, that one body warms itself, at the expense of another body which is cooled to some extent.[4] Although the transfer of heat is entirely imperceptible, there is felt-warmth or the sense-perception of warmth-condition. Very simple qualitative experiences lie behind the heat substance concept; such experiences occur in the nursery school, as it were, before the physics laboratory has been thought of. They constitute the childhood of physics, and they happen to ordinary people in the ordinary scenes of life.

[2] W., p. 183. [3] W., p. 183. [4] W., p. 185.

Mach believed that physics properly begins with such homely simple qualitative experience. Physics is an organism, something which has an evolution and has had a childhood; it is not a clever artefact which sophisticated man has created for himself. Where Mach differs from more ordinary writers on the philosophy of science is in his ability to develop cogent arguments, the fruits of an excellent historical and philosophical imagination, which most effectively demonstrate the truth of the positions he is taking. Anyone who has studied carefully the two chapters on calorimetry knows that physics is based on rough qualitative observations, and he knows why this must be true.

Richmann's formula that

$$t = \frac{m_1 t_1 + m_2 t_2}{m_1 + m_2} \tag{1}$$

can be rewritten

$$m_1(t - t_1) = m_2(t_2 - t) \tag{2}$$

where it is convenient to think of t, the temperature of the mixture, as greater than t_1 and less than t_2. If the two products in relation (2) are called 'quantities of heat' then (2) is the algebraic expression of the truth that

$$\text{heat gained} = \text{heat lost.}$$

If the Greek letter θ is used to denote 'rise in temperature' we can write

$$t - t_1 = \theta_1$$

and $$t_2 - t = -\theta_2,$$

from which $$m_1\theta_1 + m_2\theta_2 = 0 \tag{3}$$

arises as an improved form of equation (2). There is no objection to the product $m\theta$ being called a 'quantity of heat', but if we do call it such we are not committed to the philosophical view that there is such a substance as heat.[5]

Equation (3) can be generalised in order that the facts of

[5] W., p. 186.

mixing together any number of samples of water at various temperatures may be neatly epitomised:

$$\Sigma m\theta = 0. \tag{4}$$

If substances with different specific heats are considered, (4) becomes:

$$\Sigma m\theta s = 0, \tag{5}$$

where m is the mass of substance whose specific heat is s and whose temperature rise is θ. The quantity θ is like a vector in one dimension; it can be positive or negative. Mach makes his usual comment that a comprehensive relation of the type of (5) has considerable economic value, for it covers a wealth of possible cases.[6]

For two substances only, equation (5) takes the form

$$\frac{m_1 s_1}{m_2 s_2} = -\frac{\theta_2}{\theta_1}. \tag{5}$$

The product ms, called the heat capacity, can be denoted by X and (5) then becomes

$$\frac{X_1}{X_2} = -\frac{\theta_2}{\theta_1}. \tag{5'}$$

When $m_1 = m_2$, then X_1/X_2 becomes the specific heat ratio. Such mathematical expression of the main facts of calorimetry is properly scientific; the object is the abstract economic expression of the facts. No one need ask for anything more.[7]

Suppose now that samples (or bodies) with heat capacities X_1 and X_2 are mixed, and that the change in temperature for the sample (or body) whose heat capacity is X_1 is θ_1^2. The change in temperature of the second sample is represented by θ_2^1. Now consider the three experiments represented by the equations of the type (5′):

$$X_1\theta_1^2 + X_2\theta_2^1 = 0,$$
$$X_2\theta_2^3 + X_3\theta_3^2 = 0$$

and

$$X_3\theta_3^1 + X_1\theta_1^3 = 0.$$

[6] W., p. 187. [7] W., p. 188.

In fact there are here three simple equations and only two 'unknowns', viz.

$$\frac{X_2}{X_1} \quad \text{and} \quad \frac{X_3}{X_1}.$$

The three equations then *need* not all be satisfied, from the simple mathematical point of view. The experimental fact that all three equations *are* satisfied, or at least *very nearly so*, is equivalent to the discovery that

$$\frac{\theta_2^1 \cdot \theta_3^2 \cdot \theta_1^3}{\theta_1^2 \cdot \theta_2^3 \cdot \theta_3^1} = -1. \tag{6}$$

This can be generalised to any number of samples or bodies.[8] Any quotient of the type

$$\frac{\theta_b^a \cdot \theta_c^b \cdot \theta_d^c \cdot \cdot \cdot \theta_q^p}{\theta_a^b \cdot \theta_b^c \cdot \theta_c^d \cdot \cdot \cdot \theta_p^q}$$

is found to be nearly $+1$ or -1.

[8] W., p. 190.

3 Geometry and physics

1. Geometry is the physics of length

According to Kant neither physics nor mathematics rests exclusively on analytical judgements. Mach's view that geometry is essentially physics is presented without the metaphysics of Kant. It is worth consideration on its own merits.

Mach's theory of geometry is summed up in two short statements:

> . . . the art of measuring space (*die Raummesskunst*), that is, geometry . . . consists in the *comparison* of solid bodies with one another. . . .[1]
>
> But a good part of our geometry is a genuine *physics* of space.[2]

The first part of the first statement is evidently incorrect. If Mach is wrong to think that space can be seen, he is equally wrong to think it can be measured. Probably, as Kant believed, space is an *a priori* pure intuition and not an element in sense-perception. Mach's statement about the "physics of space" can be replaced by a statement about the physics of length. We have no ordinary sensuous experience of space, but we can measure the length of *a solid body* by means of a ruler or chain, the ruler or chain being *another solid body*. And this is, in Mach's own term, a *comparison*. I think one can go a little further with Mach; one can 'see length' in the same way that one can 'feel warmth' or 'see red'. Using the term 'feel' in the broad sense of Hume, what we in fact 'feel' is 'the length of a fence', 'the warmth of a cup of tea' and 'the redness of a book'. We are not obliged in the end to accept Mach's account of the 'body', or

[1] PP., p. 494. [2] W., p. 454.

'object', in its entirety; but the notion of 'the family' of sense-data (Price) remains at least a legitimate part of the notion of 'body'.

Mach requires us to consider the possibility of regarding geometry as a branch of physics. If so, one can ask: What is the 'O' concept which properly inaugurates the science of geometry? To this one can answer that it is the metrical concept of 'length', or—more accurately—that 'length' typifies three inaugurating concepts which share a common status. There is no doubt that 'length' passes the ordinal test easily. The carrying out of the test with three coins of different diameters is very likely to succeed, because, even if the pieces are held at different distances from the eye, we still tend to 'correct' the visual spatial modification of size and see them 'as they are'. This is what psychologists have called 'the perceptual constancies',[3] an interesting phenomenon discussed by Mach himself.[4] The test will not always be passed, but there is no difficulty at all in finding "some favourable examples". The following statement, which does not go much beyond Mach, is suggested as a more acceptable alternative to the one quoted from the *Principles of Heat:*

> A good part of our geometry is a genuine physics of 'length', 'area' and 'volume'; this set of metrical concepts form the generating 'O' concepts of this part of physics, just as 'force' and 'temperature' are the generating 'O' concepts of mechanics and heat respectively.

The distinction between 'seen length' and 'measured length' is valid,[5] and is represented by the two sides of the ordinal test as applied to the inaugurating concepts of geometrical science. Mach's assertion that "visual space . . . is in nowise metrical"[6] is somewhat obscure; his meaning is simply that 'seen length, area and volume' are not metrical. Even this modified statement may be too strong. Suppose we take 'the ordinal' as a realm lying between the realms of the sense-perceived and the metrical.

Sense-perceived: Ordinal: Metrical.

Our experience of things in space enables us to order them, with

[3] O. L. Zangwill, *An Introduction to Psychology*, London: 1950, pp. 29–35.
[4] P., pp. 74–5. [5] S., p. 35. [6] S., p. 7.

success in favourable instances, according to their lesser, intermediate or greater dimensions of length, area and volume. In this way, even if experience cannot be metrical, it can be ordinal. I do not see any objection to calling 'the ordinal' by some such designation as 'the rudimentary metrical'. If there were no prior ordinal experience, there would be no metrical science.

The characteristic operation of the science of geometry is measurement; it is not strictly the measurement of space ("die Raummesskunst" is Mach's designation) but the measurement of solid bodies. The essence of measurement is comparison of one thing with another; the merchant tailor of earlier times would tell out his cloth against his own arm or foot, and the modern physicist will compare the standard metre against the wave-length of a very pure species in the cadmium spectrum. The principle is the same in both cases. It is not every student of Euclid who realises that measurement is one of the essential ideas of that great system. Suppose we consider the proof of the famous 'theorem four': that if two triangles have two sides of the one equal to two sides of the other, each to each, and have also the angles contained by those sides equal; then the two triangles shall be equal in all respects.

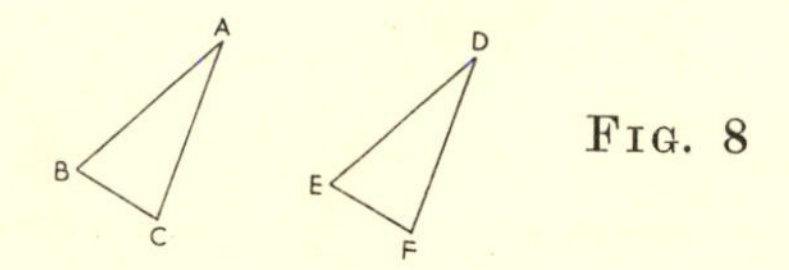

FIG. 8

The proof begins by taking up the triangle D E F and placing it on top of the triangle A B C, with D on A and the side D E along A B. Because D E = A B, then E falls on B, and so on. This is essentially the *measurement* of one triangle by the other. As a second example it may be noted that in Euclid's proof of the theorem of Pythagoras (Fig. 4) areas are indirectly 'measured off' against each other.

But is geometry, then, the physics of the diagrams in the geometry book? According to Mach this is not so very far from being the case. The triangles of 'theorem four' are derived in the first place from our experience of 'nearly plane' faces of common solid objects. The diagrams in the book require little piles of printers' ink, and the diagrams on the blackboard require little piles of chalk. Euclid's theorems are *about* ideas, ideal or perfect triangles and circles. If Mach is right, these are not Platonic transcendent realities; rather, they are 'idealised' out of homely common 'solid' things, and resemble Sadi Carnot's perfectly reversible heat engine which works on two perfectly isothermal and two other perfectly adiabatic processes. If there had never been a 'real' lever, there would never have been a 'perfect lever'; if there had never been common solid objects, there would never have been Euclid's triangles and circles.

I began the study of physics in the year 1920 and I vividly recollect one of the earliest exercises: To determine the area of an irregularly shaped piece of cardboard by two independent methods. We covered it with squared paper and counted the squares; we also weighed it, and weighed a unit area of it. The use of a balance made the exercise in mensuration *look like* physics; according to Mach it *was* physics, through and through. There was some satisfaction in reaching the same result by two such different sets of operations; and I enjoyed a vision of the purity and austerity of the great subject of physics which has remained fresh with me.

2. Three dimensions and solid bodies

We cannot imagine a position not attainable by positive and negative displacements of our body in three cardinal directions set at right angles to each other like Cartesian axes. This simple proof, that '*length*' *or* '*displacement*' *is given to us* with three dimensions, is found in Aristotle and Galileo.[1] To say that there is a group of *three* inaugurating concepts for the science of geometry is to say the same thing in a different way. Characteristically, Mach holds that a man carries in his own body that which makes him familiar with the three dimensions of vector displacement; the asymmetry of the external back and front parts of his body, the asymmetry of the external lower and upper

[1] G. J. Whitrow, 'Why Physical Space Has Three Dimensions,' *British Journal for the Philosophy of Science*, VI, no. 21, 1955, pp. 13–21.

parts, and the asymmetry of the left and right sides of the inside of his body.[2] Descartes placed his own body at the origin of his three axes of coordinates.[3] To complete this summary statement it must be recalled that Kant considered that the space of pure *a priori* intuition necessarily—not empirically—has three dimensions.

Mach discusses[4] the theorem of Möbius (1827): *A body, with maximum or general asymmetry, in a space of* n *dimensions, requires displacement in a space of* n + 1 *dimensions in order that it may be made to coincide with its mirror image.* Thus, let a b c be a rectilinear array of points, and c′ b′ a′ their mirror image in the plane mirror S S with its plane perpendicular to the plane of the paper.

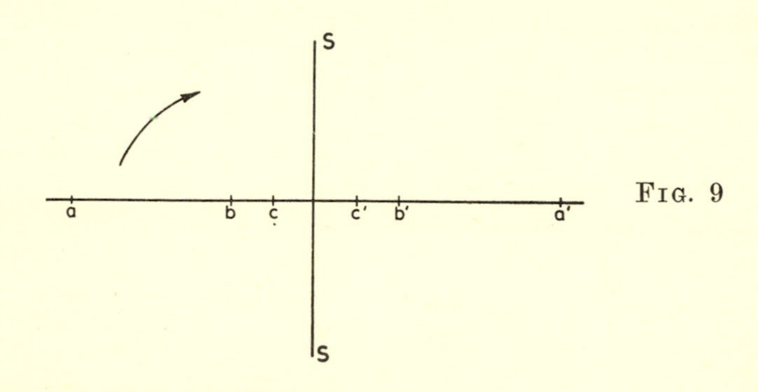

FIG. 9

We may call a b c a one-dimensional object. How, asks Möbius, can a b c be brought into coincidence with its mirror image a′ b′ c′? By movements of the one-dimensional object a b c in its one-dimensioned space a b c c′ b′ a′ this is impossible. If however we can swing a b c around in the two-dimensioned plane of the paper as shown by the curved arrow, the coincidence is achieved. It can alternatively be achieved by swinging a b c out of the plane of the paper and by a movement in the one plane at right angles to the plane of the paper, i.e. again by a movement in a two-dimensioned space. In short, to make

[2] S., p. 18.
[3] A. N. Whitehead, *Introduction to Mathematics*, London: 1939, p. 125.
[4] PP., pp. 502–4.

the one-dimensioned object coincide with its mirror image, two-dimensioned space is needed for the opreation.

The reader may like to convince himself of the general truth of Möbius' theorem by considering a plane triangle and its mirror image, and also a tetrahedron with four differently labelled vertices and its mirror image in three-dimensioned space.

A chemist may perhaps be forgiven if he is led to a naïve interpretation of the last example. The stereochemical facts discovered by van t'Hoff and others may seem to indicate that an asymmetric molecule, that of d-lactic acid for example, differs from its mirror image, the l-lactic acid molecule, for neither of these two molecules can find a fourth dimension in which to turn round. But Mach objects to high-flown speculations about the fourth dimension:

> As a matter of fact such investigations in physics have no object, as this science is occupied with what is *sensuously demonstrable*. And so there is nothing in ohe, two or four dimensions, there are only things in three dimensions. . . . The most minute physical object is three-dimensional, an element of volume, a body. Surfaces, lines and points are merely mathematical fictions.[5]

Mach insists quite rightly that the initial objects of geometrical experience must be solid bodies in three dimensions, and indeed that there are no other bodies. The d-lactic acid molecule is three-dimensional because it is a model constructed by the chemist guided in this task by the analogy of all common objects.

The world then is full of solid objects endowed with a certain stability. They can be moved about in space and they remain identical; this identity extends to their internal spatial configuration. Although in a sense areas and lengths can be measured, every measuring instrument is a solid and everything measured is solid too.[6] If we 'really' carried out Euclid's proof of 'theorem four' (Fig. 8) with the triangle D E F drawn as the mirror image of triangle A B C we should be forced to use a third dimension in order to move one of the triangles; even

[5] PP., pp. 505–6. [6] S., p. 50.

with the triangles set out in the usual way, each of them is 'really' a shallow solid prism with triangular cross-section. Mach believes that "savage tribes"[7] may have discovered that the three angles of a triangle add up to a straight angle, when they laid triangular prisms as paving stones.

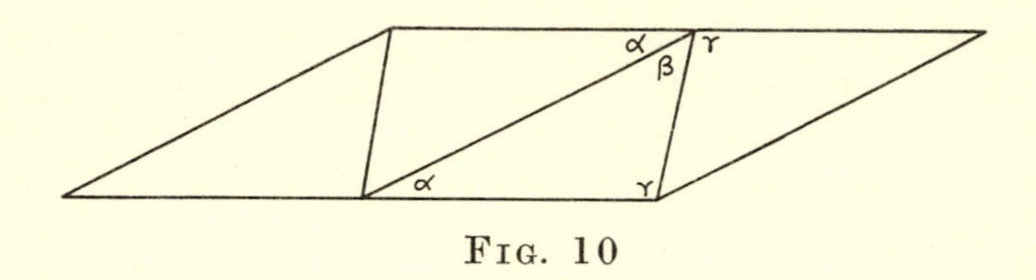

FIG. 10

Mach's comment is very bold:

> The knowledge that the angle-sum of the plane triangle is equal to a *determinate quantity*, namely, to two right angles, has thus been reached by experience, not otherwise than the law of the lever or Boyle and Mariotte's law of gases.[8]

3. Mathematics and physics compared

In Mach's view the theorems of geometry are about ideas; these ideas, as I have already noted, are not Platonic Ideas, for they are 'refined out of' experiences of the common solid objects of every-day life. The temptation to develop this line of thought into a far-reaching analogy between mathematics and physics is very strong.

Just as 'the triangle' idea is 'refined out of' a brick belonging to the savage tribes, so the perfect gas, for which Boyle's law is exactly true, is a 'refinement out of' experience with actual gases. At least some of the laws of physics are not in fact about levers, inclined planes, gases and falling bodies; they are about the *perfect* lever, the *idealised* inclined plane, the *perfect* gas and the body which falls with *absolutely* constant acceleration. The laws of geometry are not about the solid bodies we encounter in actual life; they are about the straight line, triangle, the plane circle and various kinds of idealised three-dimensional

[7] S., p. 54.   [8] S., p. 58.

constructions. If it be true that the laws and theorems of mathematics are hypothetical, then it could also be that some of the laws of physics are hypothetical.

Such views are somewhat startling, coming from one who claimed that we should be content *to describe nature*. A close study of Mach's writings shows that they emerge in his mature and later thought. Thus his comment on Stevinus, who discovered the law of the ideal inclined plane not by measurement of the forces but by reflection on the fiction of the endless frictionless chain, appears in the 1901 edition of the *Mechanics*, but not in the first (1883) edition:

> When the static relationship is rediscovered in such a manner it has a *higher* value than the result of a metrical experiment would have, which always deviates somewhat from the theoretical truth. The deviation increases with the disturbing circumstances, as with friction, and decreases with the diminution of these difficulties. The exact static relationship is reached by *idealisation* and *disregard* of these disturbing elements. It appears in the Archimedean and Stevinian procedures as an *hypothesis* without which the individual facts of experience would at once become involved in logical contradictions. Not until we have possessed this hypothesis can we by operating with the exact concepts reconstruct the facts and acquire a scientific and logical mastery of them. The lever and the inclined plane are self-created ideal objects of mechanics. These objects alone completely satisfy the logical demands *which we make of them*; the physical lever satisfies these conditions only in the measure in which it approaches the ideal lever. The natural inquirer strives to *adapt* his *ideals* to reality.[1]

In the passage which follows (1902) Mach explicitly connects geometry and physics:

> The physical metrical experiences, like all experiences forming the basis of experimental sciences, are conceptualized,—idealized. The *need* of representing the facts by simple perspicuous concepts under easy logical control, is the reason for this. Absolutely rigid, spatially invariable bodies, perfect straight lines and planes, no more exist than a perfect gas or a perfect liquid. Nevertheless, deferring the consideration of the deviations, we prefer to work, and we also work more readily, with these concepts than with others that conform

[1] M., p. 40. Italics added in accordance with German text.

more closely to the actual properties of the objects. *Theoretical* geometry does not even need to consider these deviations, in as much as it assumes objects that fulfil the requirements of the theory absolutely, just as theoretical physics does. . . . The choice of the concepts is suggested by the facts; yet, seeing that this choice is the outcome of our spontaneous reproduction of the facts in thought, some free scope is left in the matter.[2]

Mach's thesis that geometry—the mathematics of length—is a kind of physics naturally leads him to inquire whether the mathematics of number may also be a species of physics:

Unbiased psychological observation informs us, however, that the formation of the concept of number is just as much initiated by experience as the formation of geometric concepts. We must at least know that virtually *equivalent* objects exist in *multiple* and *unalterable* form before concepts of number can originate. *Experiments* in counting also play an important part in the development of arithmetic.[3]

If we think of the rise of arithmetic as an historical and psychological process, Mach's view that such a process would naturally depend on the sensuous presentation of multiple—two at least—objects, is plausible. Moreover, Descartes welded together arithmetic and geometry into a single mathematical unit. Mach's view, that if geometry is physics then arithmetic must be physics too, hardly needs further proof. In the end, Mach and Kant agree that geometry and arithmetic reveal a deep affinity, although they differ as to what this affinity precisely is.

Mach then is prepared to defend these two promises:

Physics draws its ultimate power from sensuous sources;
Mathematics draws its ultimate power from sensuous sources.

These two propositions have an undistributed 'middle term' and therefore neither of the conclusions:

Physics is mathematics (Descartes),
Mathematics is physics (Mach):

follows. One is however led to ask an important question: What

[2] S., pp. 84–6. One word of T. McCormack's translation altered.
[3] S., p. 98. Italics as in German text.

is the true difference between mathematics and physics? Here is Mach's answer in his own words:

> The *convincing power* of geometry, and of mathematics in general, does not rest on the fact that the doctrines of these subjects have been derived through some *entirely special kind* of knowledge, but merely that their experimental basis is peculiarly easy and familiar to us, that in particular this basis is frequently verified and can be verified again at any time. The province of experience of space moreover is a very much more restricted or narrow province than that of experience *in general.* The conviction that the former has been essentially exhausted readily establishes itself in our minds, . . .[4]

Briefly mathematics and physics seem different because the experiential basis of the former is relatively more simple, much more familiar, and extremely narrow in quality.

Mach's very close identification of geometry and physics through the idealised or perfect objects of both physics and mathematics is not to be denied; examples are the perfect gas in physics and the Euclidean plane triangle in mathematics. It is true also that geometry *is about* the perfect objects of the geometers, and that physics—at least in part—*is about* the perfect objects of the physicists. The object of classical geometry is the discovery of the logical attributes of the perfect or idealised objects which are roughly depicted in the geometry books. Somewhat less accurately, the object of geometry is the perfect mathematical object itself. Now the object of physics is not the perfect or idealised object in the same sense. For reasons which Mach does not discuss adequately, the physicist constructs the perfect objects. They are defined precisely, and their properties are deduced and developed logically in theoretical physics. It is also true that, because of the way in which they are defined, the physicist has the same kind of logical mastery over them which the mathematician enjoys in his proper domain. But the perfect objects can never be the *final* object of physics, although they can be the *final* object of mathematics. The difference between mathematics and physics

[4] A., p. 346. I have replaced the Williams–Waterlow translation by a more iteral one.

is therefore more profound than Mach makes out, although what he says is true and interesting. The object of physics is nature itself, and as Mach himself repeatedly tells us, physics must always be the *description of nature*, which is certainly true in the sense that physics must not be less than the description of nature. In the passage concerning the lever, Mach notes the need of the physicist to acquire logical mastery over the facts; elsewhere and repeatedly he comments on the need of the human mind to attain intellectual peace.[5] These are interesting observations, but they hardly solve the puzzle of mathematical and physical perfect objects; indeed they cannot bear the load Mach imposes upon them. We should not think highly of an external ballistician who, wishing to retain logical mastery over the facts and desiring intellectual peace, confined his attention to the flight of projectiles in vacuo in a constant gravitational field. The ideal trajectory is no doubt part of his education, part of his stock of background knowledge, but he must forever be pursuing the knowledge of the flight of actual projectiles under the given conditions. Mach is not incorrect in his view that the external ballistician needs and seeks mental peace; but this peace would be paid for too dearly were the actual facts of real trajectories disregarded. The peace is sought by the development of new mathematical methods, methods which in their complexity match the phenomena in question. The elegance of the mathematics is a secondary consideration.

If the perfect objects are not the final object of his study, why does the physicist construct and use them at all? If we consider the pressure/temperature/volume relationships of actual gases, the striking fact emerges that for many gases, over not too extreme ranges of the variables, the relationship

$$pV = rT$$

where p is the pressure,
V the volume,
T the Kelvin temperature
and r a constant for a given mass of a specific gas,

is nearly true, although never quite true. This is sufficient in

[5] A., p. 314.

itself to lead the mind to the notion of a gas which exactly obeys the relation, or to a 'perfect gas'. It is interesting too to note that the discrepancies between the behaviour of an actual gas and this 'perfect gas' depend on the specific nature of the gas, and also on the physical variables. For example, as

$$p \to 0,$$

all gases tend to 'perfection' in the defined sense. The perfect gas is a kind of standard of comparison, and is a genuine aid to study. It would surely be far-fetched to suggest that the perfect gas is constructed in the first place for the sake of our mental peace, or to give us intellectual mastery in the domain of a few simple equations with two independent variables. I suggest that in the first place the perfect gas is constructed to help in the study of actual gases. The description of these is, according to Mach, the *final* task of the physicist; in any case, it is his *proper* task. As thermodynamics progresses it is found that temperature measured by a *perfect* gas thermometer coincides with temperature measured by a *perfectly* reversible heat engine. This is an impressive achievement of theoretical physics, and is properly mathematics. This is so, because the coincidence is a logical deduction from the postulates of the perfect gas, and the postulates of the perfectly reversible heat engine. Having discovered the coincidence between the perfect gas and the perfectly reversible engine thermometric scales, when the physicist wishes to ascertain the numerical value of a temperature on this 'perfect' scale, he is obliged to take the temperature by means of a gas thermometer and then correct it in the light of the data obtained for the gas in question by such research workers as Joule and Kelvin. In fact, the Joule-Thomson effect is an accurate measurement of the 'imperfections' of each individual gas. At this stage of the proceedings the physicist is right back at the description of nature as it is, which is his proper task.

To summarise, mathematics and physics both arise out of the study of the actual world. The actual world stimulates both the mathematician and the physicist to the creation of 'perfect' or 'idealised' objects. But the 'physical objects' represent the whole fields of physics to a lesser degree than the 'mathematical objects' represent the whole field of mathematics.

Writing a year later than Mach (1903), Henri Poincaré lent his great authority to Mach's teaching. To Poincaré, as to Mach, "if . . . there were no solid bodies in nature there would be no geometry",[6] and yet the concern of geometry is not natural solids but the idealised "remote images"[7] drawn from them. P. W. Bridgman (1936) also is entirely with Mach in affirming that mathematics is an empirical science like physics.[8]

4. Non-Euclidean geometry

Mach's discussion of non-Euclidean geometries is extensive and characteristic.[1] I give here a free paraphrase of his account of the hyperbolic geometry of Lobachévski.

Euclid's 'fifth postulate' is usually expressed in these terms:

> If a straight line, falling on two other straight lines, makes the two interior angles on the same side of it together less than two right angles, these two straight lines will meet, if continually produced, on that side on which the angles are together less than two right angles.[2]

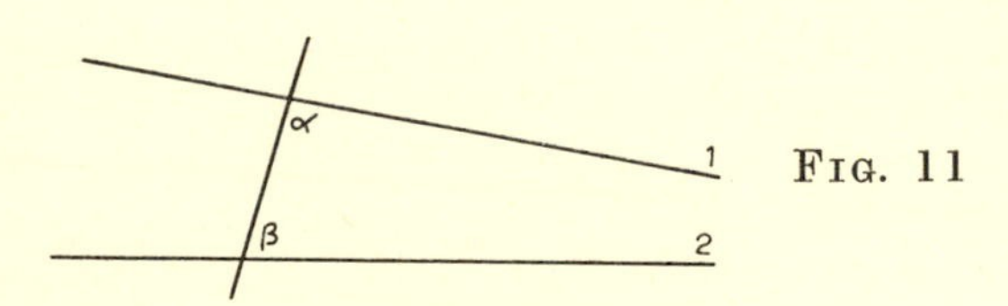

Fig. 11

Thus, in Fig. 11, if

$$\alpha + \beta < \text{two right angles},$$

eventually the lines ① and ② will meet if sufficiently produced on the right-hand side of the diagram. If this postulate is taken

[6] H. Poincaré, *Science and Hypothesis*, translated W. J. G., New York: 1952, p. 61.

[7] Ibid., p. 70.

[8] P. W. Bridgman, *The Nature of Physical Theory*, New York: 1936, pp. 52, 67.

[1] S., pp. 96–143.

[2] R. Deakin, *Euclid: Books I–IV*, London: 1897, p. 8.

for granted it follows in the first place that only one line through a point in a plane can be drawn parallel to a straight line in that plane. Many familiar and important theorems of Euclid's scheme depend on the postulate, a good example being the celebrated theorem that the three angles of any triangle add up to two right angles. Mach agrees with Professor Coxeter that the 'fifth postulate' of Euclid "is not self-evident like the others".[3] Euclid himself was reluctant to introduce it, and many mathematicians from the time of Euclid up to the time of Gauss endeavoured without success to deduce the fifth postulate from the others by logical and deductive reasoning. In 1856 Sartorius von Waltershausen published a book on the mathematical thought of Gauss. Von Waltershausen states quite clearly that Gauss came to believe that the fifth postulate cannot be deduced logically; that, on the contrary, our confidence in it depends on experience. Gauss refers to an actual experiment in which the three angles of a triangle were measured. The vertices of the triangle were at the Brocken, Hohenhagen and Inselberg and the sum of the three angles of this 'real triangle' was equal to two right angles, allowing a margin for possible error. That a mathematician of the eminence of Gauss should discuss the matter in such terms is some support for Mach's main thesis that, at its roots, geometry is experiential. One may logically substitute Euclid's fifth postulate by the theorem concerning the sum of the three angles of a triangle—this regarded as experiential; but one may not dispense with the fifth postulate altogether.

The very fact that the fifth postulate is necessarily experiential at once opens up the possibility of geometries other than the Euclidean. Let us suppose that a measurement of an angle can be made with an accuracy of not better than 1 second of arc. Suppose further that in place of Euclid's fifth postulate we make this postulate: *The sum of the three angles of a triangle is equal to 179 degrees, 59 minutes and 59½ seconds.* With respect to this postulate it could be said correctly: *It is not refuted by experience.* No doubt, this is not to say: *This postulate is verified by experience.* But, in exactly the same way, Euclid's fifth

[3] H. S. M. Coxeter, *Non-Euclidean Geometry*, Toronto: 1942, p. 2.

postulate, and the equivalent proposition about the sum of the triangle's three angles, stand in this relation to experience: *They are not falsified by it.* If it is too much to admit the error of 1 second, nevertheless, in a matter of measurement, *there must be some admitted and finite error.* Once this admission is made, that a preliminary postulate of the system is experiential, an indefinite multiplicity of geometries is also admitted. We may prefer Euclid's scheme because it seems *simpler*, but it is difficult to say in what way 180.000 is *simpler* than 179.999. In any event, it is physics we are studying, not aesthetics.

The geometry associated with the name of Lobachévski is now usually called non-Euclidean hyperbolic geometry. Mach gives a fairly extensive account of the origins of this type of geometry, and the first part of his exposition has some philosophical interest. Lobachévski's point of departure is a question: Suppose that Euclid's fifth postulate is rejected, is it possible to develop, without it, a logical scheme of geometry? To say merely that Lobachévski found that the answer to this question is in the affirmative is however somewhat misleading. What he did was to *replace* the fifth postulate by something else; his *modus operandi* was in fact similar to the scheme of the previous paragraph.

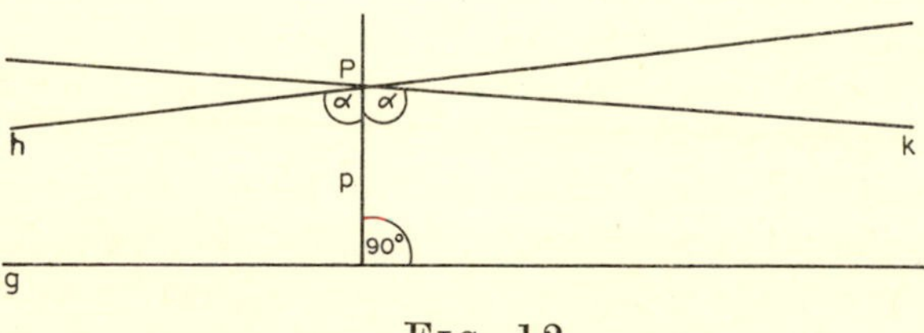

FIG. 12

In the diagram, the lower case letters g, h, k and p indicate straight lines. The capital letter P stands for a point. g is any straight line and p is another straight line perpendicular to g. P is some point on p. Through P a straight line h parallel to g is drawn; i.e. by hypothesis, it does not meet g on either side, however far the lines are produced. The angle α on the left-hand side would have to be a right angle, were Euclid's fifth postulate true. Lobachévski makes an alternative assumption, viz. that the angle α, although very nearly a right angle, is slightly less than that. *Provided the angle α is sufficiently nearly equal to a right angle, the assumption of Lobachévski is no more refuted by experience than is Euclid's fifth postulate.* Evidently if the lines g and h are conceivable in this way, another line k may be drawn making the angle α with p on the right-hand side, and this line k is also straight and parallel to g. The lines h and k make a very small, yet finite, angle with each other. Although they intersect each other, they never intersect the line g.

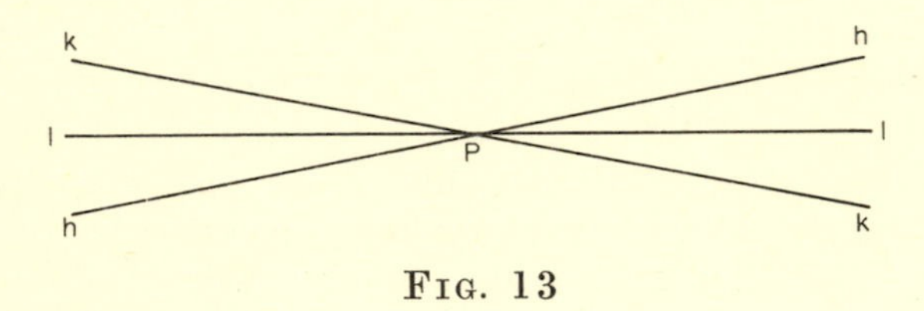

FIG. 13

Any other straight line l, which lies on both sides *within* the very small angle made by k and h together, and also passing through P, will also never intersect the line g. The lines k and h are boundaries and the two-dimensioned space is divided into two regions. One region, bounded by k and h and containing all such lines as l, is the region of lines which never intersect g. All other straight lines through P do intersect g.

Where Mach excels other writers is in the clarity of his characterisation of the *kind of thing* Lobachévski has achieved. If, in Fig. 12, the angle α were supposed to differ from a right angle by more than a very minute amount, no one in his senses would have anything to do with the new geometry. The reader can judge Mach's position for himself from the following quotations:

In the Introduction to his *New Elements of Geometry* (1835) Lobachévski proves himself a thorough natural inquirer. No one would think of attributing even to an ordinary man of sense the crude view that the "parallel-angle" [angle α, Fig. 12] was very much less than a right angle, when on slight prolongation it could be distinctly seen that they would intersect. The relations here considered admit of representation only in drawings that distort the true proportions, and we have rather to picture to ourselves that in the dimensions of the illustration [Fig. 12] the variation of α from a right angle is so small that h and k are to the eye undistinguishably coincident. . . .

Different ideas can express the facts with the *same* exactness in the domain accessible to observation. The *facts* must hence be carefully distinguished from the *intellectual* constructs the formation of which they suggested. The latter—concepts—must be *consistent* with observation, and must in addition be *logically* in accord with one another. Now these two requirements can be fulfilled in *more than one* manner, and hence the different systems of geometry.[4]

Mach is entirely consistent. Lobachévski is "a thorough natural inquirer"—a kind of physicist. But this should not surprise us, for geometry is a kind of physics.

[4] S., pp. 127, 132, 133.

4 Time

1. The time of consciousness

Mach's point of departure is that *we have a sense-perception of time*,[1] which is not obviously the case. This sense-perception, he considers, stands in ordinal correspondence with clock time, in the same way as felt warmth (Wärmeempfindung) stands with respect to temperature (Temperatur), or as 'seen length' stands with respect to 'measured length'. Therefore, to Mach, clock time, like temperature, length, force and intensity of illumination, is an 'O' concept. Yet time is not the 'inaugurating concept' of another branch of physics; it is rather a necessary condition for any kind of experience at all.

Mach's account of time is essentially in agreement with Hume who tells us that "five notes play'd on a flute give us the impression and idea of time".[2] The root of time is then a sequence of sense-perceptions or thoughts which pour into the consciousness of man. Mach calls this subjective kind of time physiological,[3] but I consider the term—'the time of consciousness'—preferable. I hear the first note of the flute. My consciousness feels the flow of time as I hear the second note; I feel the third note as after the second, and so on. So Mach boldly declares that I have a sense-perception of time (eine Zeitempfindung). Evidently, an ordinal scale could be set up by anyone, to mark out the ordered occasions of the past time of his consciousness. Mach appears to have had in mind acts of sense-perception, or experiences which can be broken down into such acts. I do not see why 'having a thought' should not count for

[1] A., p. 8.

[2] D. Hume, *A Treatise of Human Nature,* edited L. A. Selby-Bigge, Oxford: 1896, p. 1.

[3] E., pp. 423, 424.

a position on the scale. A clock of the time of *my* consciousness might then be like this:

1. I heard a church bell toll;
2. I reflected on the probable brevity of my life;
3. I looked at the road and saw three people walking;
4. I heard the church bell toll again;
5. I began to walk;
6. I felt a pain at my heart; and

NOW.

Although this interpretation of the time of consciousness may be correct as far as it goes, Mach considers that it is incomplete.[4] For example, the ordering of the miscellaneous sense-perceptions and acts of thought, by numbers, is powerless to account for our sense of the *pace of time*. Why should the slow movement of Mozart's Symphony in G minor sound slow, and the Finale of the same work sound fast? Even more puzzling, why should these movements be recollected or remembered by the lover of music as slow and fast respectively?[5]

2. Irreversible change

Mach considers that we become aware of time as irreversible, because there are in the natural world irreversible processes. It is not that every event is clearly irreversible; the idea of the irreversibility of time is conveyed to us when we find that *not all* nature can be reversed by us.

> In so far as a portion only of the changes of nature depends on us and can be reversed by us, does time appear to us irreversible, and the time that is past irrevocably gone.
>
> We arrive at the idea of time—to express it briefly and popularly—by the connection of that which is contained in the province of our memory with that which is contained in the province of our sense-perception. When we say that time flows on in a definite direction or sense, we mean that physical events generally (and therefore also physiological events) take place in a definite sense.

[4] E., p. 427.

[5] J. Bradley, *Ernst Mach's Philosophy of Science*, Ph.D. Thesis, University of London, 1965, pp. 617, 618.

Differences of temperature, electrical (potential) differences, differences of level generally, if left to themselves, all grow less and not greater. If we contemplate two bodies of different temperatures, put in contact and left wholly to themselves, we shall find that it is possible only for greater differences of temperature in the field of memory to exist with lesser ones in the field of sense-perception, and not the reverse.[1]

So Mach would have us believe that progress of 'felt time' is indicated to us not only by a spontaneous diminution of temperature difference, but by the diminution of differences in general. It is as if the correlation between time and diminishing temperature difference, this taken exclusively, fails to satisfy the philosophic mind: " . . . an uncompensated spontaneous enhancement of a difference does not happen."[2]

As a physicist, Mach is obliged to inquire into what precisely is meant by the assertion that we can reverse the changes of nature in part only, or to a limited extent. The truth seems obvious enough for such important changes as the birth and death of an individual. But in other cases, there is some ambiguity. Left to themselves the hot body cools off and the colder body touching it warms up; yet we can heat up the former body again by putting it in the fire and repeat the experience. In fact, if words are used in the ordinary way, the process is reversible. Yet since 1824, the date of Sadi Carnot's monograph, physicists have learned to regard this simple change as the very type and exemplar of the irreversible process. The metrical concept of entropy is required to reconcile the simple every-day facts with the technical idea of irreversibility. In both the spontaneous cooling of the hotter body, and again in the artificial restoration of the high temperature to this same body, there are over-all increases in entropy. To make this judgement, one has not only to consider the original two bodies; one has to take into account the fire used, and anything else in the universe which undergoes any kind of change. The concept of entropy is a clever mode of estimation of the element of irreversibility in all changes whatever. Even the swing of a pendulum has an irreversible 'component' in it. When the

[1] M., pp. 274–5. [2] E., p. 436.

pendulum swings from left to right and then again from right to left,[3] the irreversibility of all actual happenings mounts up systematically by two increments. *Whatever actually happens* causes an increase in entropy.

. . the entropy of the universe, if it could ever possibly be determined, would actually represent a species of absolute measure of time.[4]

It is interesting to compare this modest statement with Eddington's fanciful idea of a man having an entropy clock in his brain in order that the notion of time may somehow be conveyed to his mind.[5]

Mach's reference to entropy and "a species of absolute measure of time" suggests that if the absolute time referred to by Newton in the *Principia* could have any meaning at all, it would be found in the identification of absolute time with the time of a clock 'telling the total entropy of the cosmos'. It could be that the only clock which cannot be made to go backwards is the 'total entropy' clock, although strictly this is not known to be true and the assumption of its truth represents a great extrapolation beyond our experience. The time of 'the total entropy clock' could also be the time of natural history; in which kind of time there was a time when there were invertebrates on earth but not vertebrates, a later time when there were also fishes but as yet no mammals, and so on. But since the total entropy clock and the person who tells the time by it are both imaginary, the idea that entropy is time's arrow is a useless metaphysical speculation.

3. Two orders of time and temperature

As I have indicated in the previous chapter Mach's view that "spaces and times may just as appropriately be called sensations as colours and sounds",[1] taken together with his strange notion that the elements of time and space are easily detachable from the 'molecules' of experience, is in need of Kantian revision and

[3] E., p. 436. [4] M., p. 276.

[5] A. S. Eddington, *The Nature of the Physical World*, Cambridge: 1930, p. 101.

[1] A., p. 8.

reformulation; but it has fruitful consequences, the most valuable of which is an analogy between the two 'O' concepts, time and temperature.

Prima facie one might think it hard to believe in a specific sense-perception of time, because man has no specific sense organ for this specific kind of sense-perception. Mach is not troubled by this, and a little reflection shows that he is justified. For, although a man distinguishes colours by his eyes, he needs no specific sense-organ in order to distinguish two warmth-conditions. Any part of his body will give him a warmth-sense-perception (Wärmeempfindung) of a warmth-condition (Wärmezustand). Mach makes the interesting suggestion that *we become aware of time because of the work of attention, and the organic consumption which is a kind of index of the labour of our waking hours*. In Mach's own words:

> Many sensations make their appearance with, others without, a clear sensation of space. But time-sensation accompanies *every* other sensation, and can be wholly separated from none. . . .
> That a definite, *specific* time-sensation exists, appears to me beyond all doubt. . . .
> Since, so long as we are conscious, time-sensation is always present, it is probable that it is connected with the organic *consumption* necessarily associated with consciousness,—that we feel the *work of attention* as time. . . .
> The fatiguing of the organ of consciousness goes on continually in working hours, and the labour of attention increases just as continuously. The sensations connected with *greater* expenditure of attention appear to us to happen *later*.[2]

By 'work' Mach means physical work measured in ergs; the "work of attention" may be taken to include the work done by the human body in its metabolic processes.

That man has no specific sense-organ for warmth (Wärmeempfindung) and no specific sense-organ for felt time (Zeitsinn) constitutes a negative analogy between temperature and time. Positively, if Mach's view that the time of consciousness is a sense-perception can be accepted, then metrical time (clock or physical time) is an 'O' concept, just as temperature is. There is no difficulty in finding ordinal agreement between three

[2] A., pp. 245, 248, 250–1.

feelings of time and three *measurements* of time in a favourable case. Had we asked Mach whether time, like force and temperature, is a fact (eine Tatsache), or whether it is an addition (eine Zutat) to other facts, he would certainly have replied that time also is fact. His interesting comparison of time with temperature is evidence in favour of this interpretation of his thought. Time then, according to Mach, is an 'O' concept. The pre-concept or shadow-concept 'behind it' is the time of consciousness. The following table and two quotations summarise Mach's doctrine of temperature/time:

Felt warmth (Wärmeempfindung)	Felt time (Zeitsinn)
Temperature (Temperatur)	Clock time (Physikalische Zeit)

In the ideas of time, the *sense-perception of duration* vis-à-vis different time measurements plays the same role, as the sense-perception of warmth in the above case [plays vis-à-vis temperature].[3]

One must distinguish just as strictly between the immediate sense-perception of a duration and a metrical time interval, as between the sense-perception of warmth and temperature.[4]

4. Clock time

In principle, the simplest kind of clock is the inertia clock, which however Mach does not describe.[1] This consists of a geometrical straight line in remote space, along which a particle is moving according to Galileo's law of inertia. The line is uniformly graduated in length units; the 'time' is simply the length reading of the particle. Reference to this clock is included here, because thereby the remarkable analogy between time and temperature is further strengthened. In principle, reading of thermometer and reading of clock can both be lengths.

Astronomical or sidereal time is defined and measured in terms of the angular spin of the earth, the rotation of the earth about the axis of spin being judged with respect to the stars. Just as when a temperature is read by a thermometer, the sense-perception of a length (of a mercury thread, for example) is

[3] W., p. 52. [4] E., p. 433.

[1] H. Dingle, *The Special Theory of Relativity*, London: 1961, pp. 37–9.

substituted for the primary sense-perception of warmth (die Wärmeempfindung); so when physical time is recorded as an angular displacement, the sense-perception of the rotation of the angle arm, easily reduced to clock pointer readings and closely akin to the sense-perception of a length, is substituted for the primary sense-perception of the time of consciousness (der Zeitsinn). In both cases, the transition from what I have called the shadow-concept to the metrical concept proper, does not entail the complete elimination of sense-perception; indeed the transition does not involve a great deal more than the replacement of Hume's *heard* flute sounds by the astronomer's *seen* spin of the earth.

Mach's first important book, the *Conservation of Energy*, contains a formal statement of his view of time.[2] Let ϕ_1 represent the metrical aspect of some phenomenon, e.g. the length of fall of a particle in the earth's gravitational field. Then,

$$\phi_1 = f_1(\tau) \tag{1}$$

where f_1 means 'some function of' and τ = astronomical or clock time.

Although τ means physical time, by an agreed convention, it represents in fact an angular displacement of the earth, a number of swings of a pendulum or a length measurement. The equation (1), which looks like a relationship between a lapse of time and a physical phenomenon, is in fact *a relationship between two physical phenomena*. Let us consider a second relationship of the same kind and form, to which Mach's critical commentary applies in exactly the same way:

$$\phi_2 = f_2(\tau). \tag{2}$$

ϕ_2 may be, for example, a number of units of heat conducted during the time τ. From (1) and (2) it will follow that

$$\phi_1 = \psi(\phi_2) \tag{3}$$

where ψ means 'some function of'. Mach's point is that all three equations represent the same kind of thing. But he admits that the derivation of (3) from (1) and (2) suggests that the concept

[2] GG., pp. 35, 57.

τ is a kind of common currency, or a medium of exchange, between the metrical aspects of physical phenomena.

The earth's angle of rotation is very ready to our hand, and thus we easily substitute it for other phenomena which are connected with it but less accessible to us; it is a kind of money which we spend to avoid the inconvenient trading with phenomena, so that the proverb "Time is money" has also here a meaning. We can eliminate time from every law of nature by putting in its place a phenomenon dependent on the earth's angle of rotation. . . . Physically, then, time is the representability of any phenomenon as a function of any other one.[3]

Mach remained faithful to this doctrine of time. In his last *Popular Lecture*,[4] added to the fourth edition in 1910, he refers to physical time as "mutual interdependence of changes".[5] Thus, whilst the earth turns 1/86 400 of a revolution about its axis of spin, light travels 300 000 km, Newton's apple falls 4.9 metres, a simple pendulum of length 1 metre performs a single oscillation, and so on. If time is defined in terms of the rotation of the earth about its axis of spin, there is no puzzle at all concerning the proportionality between time and the angular position of the earth. It is however perhaps puzzling that, *in a manifold way*, several other changes in the universe 'keep step' with time, so defined.

Is not this very extraordinary and singular? What can all these processes have to do with the rotation of the earth?[6]

5. Mach's temperature clock

In *Knowledge and Error* Mach develops a novel kind of clock,[1] based on the idea that temperature is *philosophically* preferable to angular displacement as a conceptual basis on which to erect a metrical time scale. In the exposition of this 'intellectual experiment', I have attempted to clarify the form and style of Mach's account in unessential details.

The diagram (Fig. 14)[2] represents a plane section of a solid cylinder whose axis is at right angles to the plane of the paper.

[3] Ibid. [4] PP., pp. 492–508. [5] PP., p. 495. [6] Ibid.
[1] E., pp. 434–9. [2] E., Fig. 34.

The solid is made up of three equal 'sectors' in contact as shown. Each of the three solid sector bodies is endowed with infinitely good internal thermal conductivity.

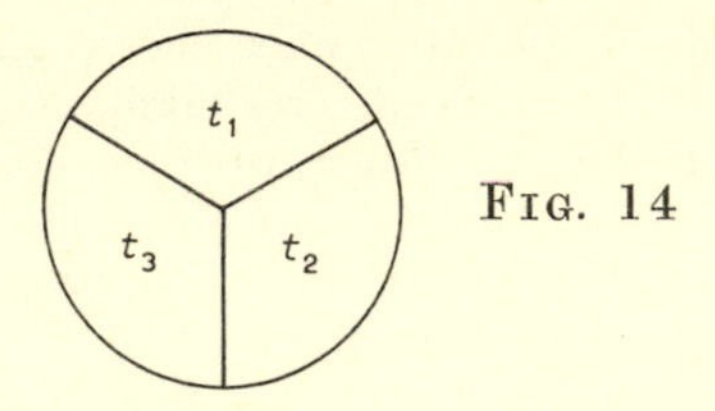

FIG. 14

There is no question therefore of two parts of a single sector body being at different temperatures. Between each of the three pairs of equal sector bodies, there is the same area of contact. Further, one and the same coefficient of external conductivity of heat obtains between each of the three pairs; in other words, all three sector bodies have the same Newtonian cooling characteristics. It is also assumed that all three sector bodies are made of the same substance, a kind of ideal super-metal, with the same specific heat, and that this specific heat is invariant with temperature. The three sector bodies have equal masses, and they neither expand nor contract with changes of temperature. Finally, the whole cylinder is supposed to be thermally insulated from the rest of the universe; no exchange of heat, other than exchanges between the sector bodies, can possibly occur.

Now time begins, as it were. At time

$$\tau = 0,$$

the initial temperatures of the three sector bodies are (say) T_1, T_2 and T_3, where

$$T_1 > T_2 > T_3.$$

The quantities T may be thought of as Kelvin or 'absolute' temperatures. At any subsequent time τ (variable) the three temperatures T_1, T_2 and T_3 have become t_1, t_2 and t_3, where t_1, t_2 and t_3, like τ, are variables. Because of the exclusion of the cylinder from the external universe,

$$t_1 + t_2 + t_3 = T_1 + T_2 + T_3 = c, \qquad (1)$$

where c is some constant with the dimensions of temperature.

We now write an equation for the change of t_1 at any time τ, in accordance with Newton's law of cooling.[3]

$$\frac{dt_1}{d\tau} = k(t_3 - t_1) + k(t_2 - t_1), \tag{2}$$

where k is a Newtonian cooling constant.

Since $$t_3 < t_1$$
and $$t_2 < t_1,$$

both right-hand members of equation (2) are negative. This simply means that

$$\frac{dt_1}{d\tau} \text{ is negative,}$$

or that the hottest of the three sector bodies cools. Equations (1) and (2) taken together give

$$\frac{dt_1}{d\tau} = k(c - 3t_1). \tag{3}$$

The corresponding equations for the other two sector bodies are

$$\frac{dt_2}{d\tau} = k(c - 3t_2) \tag{3b}$$

and

$$\frac{dt_3}{d\tau} = k(c - 3t_3). \tag{3c}$$

Evidently,

$$c - 3t_1$$

must be negative, and

$$c - 3t_3$$

must be positive. On integration, equation (3) yields

$$c - 3t_1 = K \exp(-3k\tau) \tag{4}$$

[3] Mach interprets Newton's law in a very wide sense. There is some historical justification for this wide interpretation.

where K is the constant of integration. K is easily evaluated from

$$c - 3T_1 = K,$$

where $$\tau = 0.$$

In this way, the three equations

$$\frac{c}{3} - t_1 = \left(\frac{c}{3} - T_1\right) \exp\,(-3k\tau) \tag{5}$$

$$\frac{c}{3} - t_2 = \left(\frac{c}{3} - T_2\right) \exp\,(-3k\tau) \tag{5b}$$

and

$$\frac{c}{3} - t_3 = \left(\frac{c}{3} - T_3\right) \exp\,(-3k\tau) \tag{5c}$$

are derived.

The only two variables in equation (5) are t_1 and τ and one can in principle 'tell the time' by reading the thermometer which registers t_1 instead of reading the heavens, or a good pendulum clock tuned in to the mode of the earth's spin. The temperature clock would never need winding up, because only as

$$\tau \rightarrow \infty$$

does

$$t_1 \rightarrow \frac{c}{3} \;(= t_2 = t_3 \text{ at } \tau,\ \infty).$$

In practice the three sector bodies would become sensibly at the same temperature very soon; in theory, they would never actually attain the same temperature. It is an obvious consequence of Newton's law of cooling that two bodies at different temperatures, in thermal contact, can never attain the same temperature.

The equations (5) can be expressed more elegantly by writing

$$v_1 \equiv \frac{c}{3} - t_1$$

and $$V_1 \equiv \frac{c}{3} - T_1,$$

so that $$v_1 = V_1 \exp\,(-3k\tau). \tag{6}$$

From the two corresponding equations (6b) and (6c), which it is unnecessary to write down, it also follows that

$$\frac{v_1}{V_1} = \frac{v_2}{V_2} = \frac{v_3}{V_3} \tag{7}$$

at any given time τ. Finally, (6) can be transformed to

$$\tau = \frac{1}{3k} \ln \frac{V_1}{v_1}, \tag{8}$$

and again corresponding equations (8b) and (8c) follow from equation (7).

τ in equation (8) is neither the 'time of consciousness', nor is it Newton's 'absolute time'; it is merely an angular rotation. To remind ourselves of this, we write

$$\tau \equiv \theta$$

and rewrite equation (8) in the form

$$\theta = \frac{1}{3k} \cdot \ln \frac{V_1}{v_1} \tag{9}$$

in which the variables are θ (an angle) and v_1 (a temperature difference). From (9), it is evident that to call θ time, rather than v_1 or $\ln (V_1/v_1)$ is arbitrary. If in fact we decide to take $\ln (V_1/v_1)$ as time, we get the same kind of time scale as the familiar astronomical one, but the unit of time is different. In order to 'tell the time' we look at a thermometer instead of looking at the stars.[4] Equally we may decide to take

$$\frac{V_1}{v_1} \equiv \tau'$$

as time. τ' is also read from a thermometer.

This covers the main points of Mach's own discussion. When the τ' time scale is used, some of the metrical laws of physics take on a new mathematical form. For example the fall s of a body in the earth's constant gravitational field, occurring during the time interval τ', is given by

$$\tau' = \exp (A\sqrt{s})$$

where A is a constant.[5] As Mach insists, the rôle of mathematics

[4] What follows is a slight addition to Mach's statement.

[5] The proof of this is left to the reader.

in physics is entirely secondary. *Physics is not applied mathematics.*

6. Newton's absolute space and time

In order to validate his system of dynamics Newton required "absolute, true and mathematical time" flowing "uniformly on, without regard to anything external", together with "absolute space" which "always remains similar and immovable".[1] It is difficult to see what precise meaning can be given to the term *f*, 'the acceleration of a particle of mass *m*', if one cannot have these absolute entities. Nor does Kant's account of space and time provide viable alternatives.

Yet we cannot have them. Mach's rejection of Newtonian time and space led him at last to a significant reconstruction and an extensive criticism of Newton's dynamics. The following words are the beginning of this critical movement:

> It would appear as though Newton in the remarks here cited still stood under the influence of the mediaeval philosophy, as though he had grown unfaithful to his resolves to investigate only *actual facts*. When we say a thing A changes with time, we mean simply that the conditions which determine a thing A depend on the conditions that determine another thing B. The vibrations of a pendulum take place *in time* when its excursion *depends* on the position of the earth. Since, however, in the observation of the pendulum, we are not under the necessity of taking into account its dependence on the position of the earth, but may compare it with any other thing (the conditions of which of course also depend on the position of the earth), the illusory notion easily arises that *all* the things with which we compare it are unessential. . . . This absolute time can be measured by comparison with no motion; it has therefore neither a practical nor a scientific value; and no one is justified in saying he knows aught about it. It is an idle metaphysical conception.[2]

7. Conclusion

In the last edition of the *Mechanics* only four years before his death, Mach adds these words on astronomical time:

[1] M., pp. 272, 276. Quotations from *Principia*, translated T. McCormack.
[2] M., pp. 272, 273.

We measure time by the angle of rotation of the earth, but could measure it just as well by the angle of rotation of any other planet. But, on that account, we would not believe that the *temporal* course of all physical phenomena would have to be disturbed if the earth or the distant planet referred to should suddenly experience an abrupt variation of angular velocity.[1]

This is so, because astronomical time is not absolute; it is judged by the rotation of the earth *relative* to the remote stars. As for all clocks which work on the displacement principle, the displacements in question are relative; absolute displacement is meaningless. Hence a sudden change in the angular velocity of spin of the earth would be no more significant than a sudden increase in length of the pendulum of a good pendulum clock; in both cases, the *one* change would only be judged in terms of the *many* other absences of change. The astronomical clock is no more than *primus inter pares*.

Are we permitted to imagine that all relative velocities in the universe could suddenly change in such a way that all 'relative velocity' (displacement) clocks remain in step, and therefore in such a way that we could not know of the change? As a thought-experiment, this is a non-starter; the change would have to be a digression from absolute Newtonian time which has been rejected.

We are left with Mach's extraordinary and singular fact; displacement and temperature clocks keep in step. In more ordinary language, a pure substance such as copper keeps the same coefficient of thermal conductivity which is a function of displacement clock time. What is constant in nature is a kind of ratio; one is reminded of special relativity theory, in which the *relative* velocity of light is *constant*. It is well worth while from time to time to redetermine, with the greatest possible precision, such physical constants as thermal conductivity coefficients; an authentic drift in value of such a constant would be significant.

[1] M., p. 295.

5 Mach and dynamics

1. Introduction

Mach's account of the dynamics of Newton sharply focuses his teaching on the two main kinds of metrical concept; for 'force' —which 'sets mechanics off' as Bridgman would put it—is an 'O' concept, whereas 'mass' is a 'non-O' concept quite remote from sensuous experience. The *Mechanics* is at once critical and historical; the first phase of Mach's criticism of mechanical theory is within the classical tradition of Newton, but the second phase goes beyond this tradition and approaches—so Einstein judged—relativity physics. These two phases correspond roughly to this chapter and its successor.

The discussion of these themes will enable us to extend the treatment of the concept of temperature, and to follow Mach in his critical account of the conservation of energy.

2. The concept of force: the foundations laid by Galileo

The sciences of statics and dynamics as they developed in history ". . . contain a double mode of interpretation of 'force': on the one hand force as a push or pull, on the other hand force as a circumstance determinative of acceleration."[1] The position is logically equivalent to having *two* concepts of force, which *a priori* should not be judged to be identical (Bridgman). These two concepts require reconciliation. One example of this reconciliation has been given in connection with the discussion of Hooke's theory of 'return'. No contradiction arises from assuming that one force on the lump of lead may be judged *statically* from the extension of a spiral spring,

[1] E., p. 174.

and that another force on the lump of lead may be judged *dynamically* from the product of its mass and centripetal acceleration.

The basis of 'force' as 'acceleration-determining' is found in Galileo's experiments on the 'law of equal heights'.[2] Galileo made very smooth wooden double inclined planes, and allowed a ball to roll down from one side and up the other (Fig. 15).

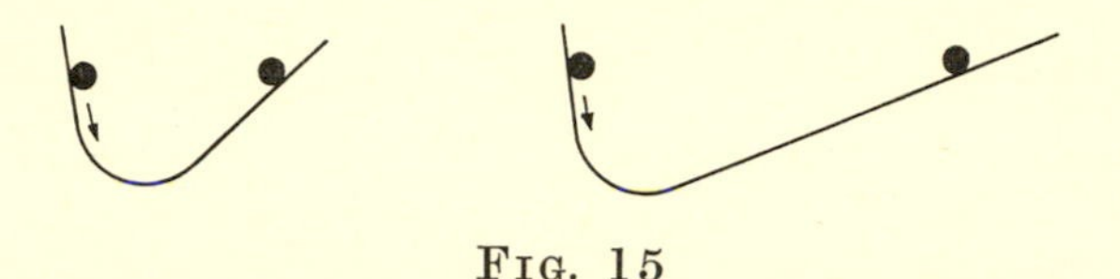

FIG. 15

He discovered that the ball by virtue of the motion acquired at the lowest point of its journey could rise, or more strictly *very nearly* rise, to the same height as that from which it had fallen; but that, on no account, could the ball rise higher than its level of origin. The law of equal heights is exhibited in the diagram. With remarkable physical insight, Galileo compared these facts with another set of facts exhibited in the swing of a pendulum (Fig. 16).

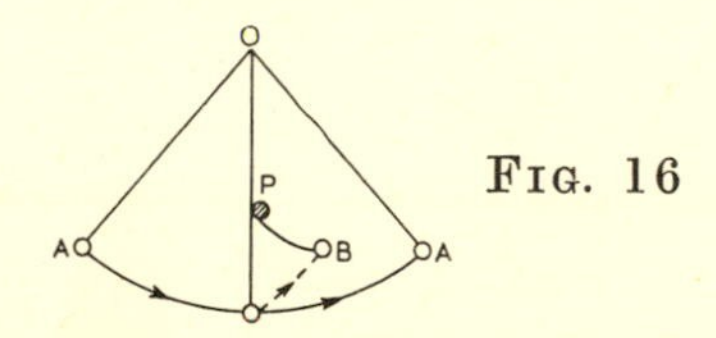

FIG. 16

The lines OA show the bob at the extremities of its excursion with AA at equal heights. If the pencil P is placed in such a way that the bob is forced to climb upwards by a new path, nevertheless the height B attained (Fig. 16) is the same as before. Even the instructive case, when P is held so low that the bob cannot recover its original height, and, instead, the bob wraps itself around the pencil, is discussed by Galileo. I think the

[2] M., pp. 162–70. G., pp. 23–8. E., pp. 135–40. G. Galilei, *Dialogues concerning Two New Sciences*, translated H. Crew and A. de Salvio, New York: 1914, pp. 215–18, pp. 170–2, p. 244.

analogy between the two sets of experiments is perhaps the most beautiful in the history of physics.

Returning to the first experiment (Fig. 15) Galileo had the audacity to inquire: What would happen to the ball if the incline on the right-hand side were to become perfectly horizontal and perfectly smooth? Evidently the ball would continue in a straight line at unimpaired speed indefinitely, for it could never recapture its lost height (Fig. 17).

FIG. 17

Mach's critical commentary on Galileo's discovery of the law of inertia is here shortened and modified in unessential details. By an extraordinary mechanical intuition Galileo felt quite correctly that the tension in the string (Fig. 16) plays the same part as the normal reaction (Fig. 15). He also extrapolated the system of Fig. 15 to the system Fig. 17, this part of his project being correctly called a 'thought-experiment' (Gedankenexperiment). Galileo's 'thought-experiment' retains the rolling of the ball and the normal reaction of the smooth horizontal plane upon it; Newton further idealises the system so that in his formulation of the first law of motion the 'ball' is replaced by some nondescript 'body' and the plane for the motion's support is withdrawn altogether. In the end we are left with a 'perfect' or 'ideal' system, like the 'perfect gas' and the 'ideal lever' previously mentioned. Yet we must not think, as Descartes did, that the law of inertia is 'necessary' and *a priori*. It is—I am sure Mach is correct—an empirical principle or experimental generalisation. To assert the law as general requires an act of induction; the induction is applied to the 'thought-experiment' which is generalised. The processes of *abstraction*—the removal of the spin of the ball and the removal of the smooth horizontal plane—and *idealisation*—the transition from the very smooth surface to the perfectly smooth surface—coalesce as it were in the single *thought-experiment*. After the *induction* to the general principle, the 'first law' is applied with great boldness; how, for example, could the kinetic theory of gases stand up to critical inspection if one were obliged to ask: What force keeps the gas molecule moving?

Reflection on the experiments with the double incline and pendulum led Galileo to the important conclusion that a body acquires the same speed when it falls through the same vertical height, independently of the route taken. In particular, it acquires the same speed when it falls freely and in a vertical line. It is not necessary to go into the details of his most celebrated experiment on the free fall of a body. The body is accelerated according to the simple law

$$v \propto t$$

although what is tested in the experiment is the calculated consequence of this hypothesis, viz.

$$s \propto t^2,$$

where t = time of fall,
v = speed at t,
s = distance fallen in t.

Galileo was thus able to show that, when a body is isolated from force, it preserves its velocity unimpaired and moves with zero acceleration; when it falls under its own weight, it has constant acceleration downwards.

It is surprising perhaps that, having gone so far, Galileo did not take the further and decisive step: an accelerated movement is a sign of force. In fact this step was taken not by Galileo, but by Newton. For example, Galileo never inferred the existence of a central force between moon and earth, or between the sun and the various planets. He remained Aristotelian in the sense that he was disinclined to apply the terrestrial rule of uniform rectilinear motion to celestial motions. He was able however to solve delle Colombe's[3] problem: Why should a weight, allowed to fall from a high tower, land at the base of the tower, if the tower, according to Copernicus, is moving laterally during the fall? Galileo realised that, by the inertia principle, any lateral motion of the tower must be shared by the falling weight. The solution of delle Colombe's puzzle is, I believe, the first hint of

[3] F. Sherwood Taylor, *Galileo and the Freedom of Thought*, London: 1938, p. 127.

a unified mechanics in which *a single set* of laws of motion is employed. Aristotle's physics was *dualist*, with separate laws of motion for celestial and terrestrial events.

Because the velocity of a falling stone varies continuously, Galileo was obliged to *think* of the relationship between s, t, and v as

$$v = \lim_{\delta t \to 0} \frac{\delta s}{\delta t}$$

although he could not use this language. Indeed, he expresses the philosophy of the calculus in words written some 27 years before Newton differentiated a simple algebraic function:

> I hardly think you will refuse to grant that the gain of speed of the stone falling from rest follows the same sequence as the diminution and loss of this same speed when, by some impelling force, the stone is thrown to its former elevation: but even if you do not grant this, I do not see how you can doubt that the ascending stone, diminishing in speed, must before coming to rest pass through every possible degree of slowness . . . a heavy rising body does not remain for any length of time at any given degree of velocity. . . .[4]

To Aristotle there is one explanation[5] why a stone rises vertically when thrown, and another explanation of its return to the ground; to Galileo, the whole motion is characterised by a constant acceleration which the stone has when it is rising, when it is at rest at the highest point, and when it is returning.

3. Newton's third law of motion

The third law of Newton unites the 'push-pull' and 'acceleration-determining' views of force. The two views correspond in fact to the two branches of the subject, statics and dynamics. Suppose a piece of plasticine is weighed on a spring balance, and the weight is 100 g weight.

Suppose further that the plasticine is sliced into two pieces and returned to the force-meter. The weight is still 100 g wt. Evidently a whole set of forces of the 'push-pull' type is removed

[4] G. Galilei, *Two New Sciences*, pp. 164–5. [5] M., p. 171.

by slicing the plasticine, and if P stands for such a force, the experiment proves that

$$\Sigma P = 0.$$

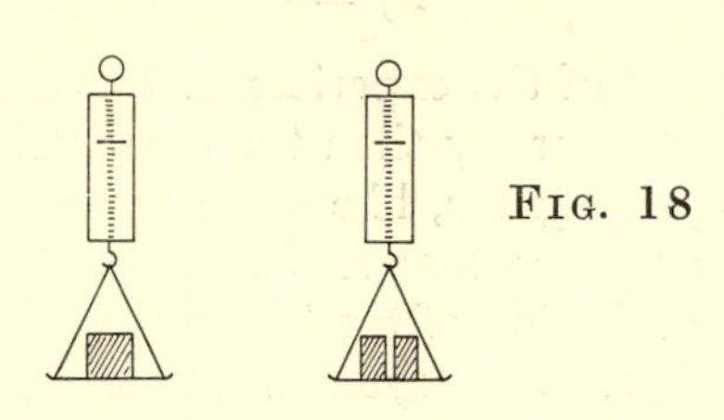

FIG. 18

Mach correctly takes the third law as experiential; that which is *experiential* can be made to appear *experimental* by the rather artificial little experiment just described.

An experiment to illustrate the dynamic application of the third law is described with parsimony by Newton,[1] and noticed briefly by Mach.[2] I demonstrate the experiment to my students in this way. Two equal powerful steel magnets are supported

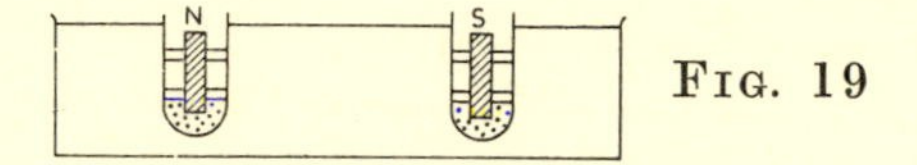

FIG. 19

vertically in large glass test-tubes weighted suitably with lead shot so that they float vertically in water as shown. The N-seeking pole of the one is opposite the S-seeking pole of the other. The tubes, held apart by hand, are released together. They move towards one another, at first slowly, then ever more and more rapidly, until collision, when they promptly come to rest. Evidently the forces, exerted by the magnets mutually, set up the opposite accelerations, and both the forces and the accelerations are equal numerically but opposite in sign, throughout the experiment and just before collision. Using P

[1] I. Newton, *Mathematical Principles*, translated A. Motte, edited F. Cajori, California: 1934, pp. 25–6.
[2] M., p. 245.

and f for force and acceleration respectively, equations of the type

$$\Sigma P = 0$$

and

$$\Sigma f = 0$$

hold at each instant of the experiment. But one of these equations is true more generally than the other. The experiment is repeated with two magnets, like poles bound together, in one tube; and with only one magnet in the other tube, as before (Fig. 20). The acceleration of the larger system is, *at each stage*, less than that of the smaller system. On collision, the two tubes come to rest as before.

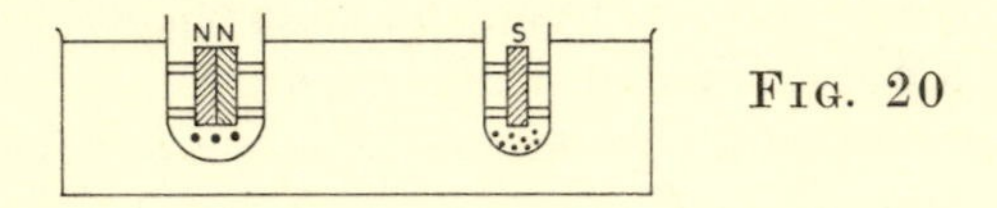

FIG. 20

The equation,

$$\Sigma P = 0$$

is still true, but this time

$$\Sigma f \neq 0.$$

The important consequence, made quite clear by this experiment, is that, *for both static and dynamic systems*, the equation

$$\Sigma P = 0$$

holds.

It has already been noted that Galileo did not take the vital step: an accelerated motion is a sign of force; and it has been stated that Newton did take that step. Newton's magnet experiment, developed in the way suggested here, shows very precisely what exactly is the position reached by Newton. In the second part of the experiment the forces on the two tubes are equal and opposite—say, at any instant, P and $-P$. But these forces do not produce in the tubes equal and opposite accelerations. Let us suppose that the accelerations are f_1 and $-f_2$ respectively. The items P, f_1 and f_2 are conventionally taken as positive. Accelerating force cannot be measured by the acceleration to which it gives rise, because, by experiment, f_1 and f_2 are not equal. It could be that

$$\text{force} \propto \text{acceleration}$$

or

$$P \propto f;$$

but if so, in the same example,

$$P = m_1 f_1$$

and

$$P = m_2 f_2$$

where m_1 and m_2 are terms multiplying the accelerations.

In the same way two very dissimilar weights W_1, W_2 can be measured statically using a Hooke's law force-meter, or spring balance. In the use of this instrument, the spring exerts an equal but opposite force upwards on the object being weighed. This is in accord with Newton's third law, in its statical application. When we think of force as 'acceleration-determining' and recollect that weight is a species of force we soon discover that we cannot estimate the values of the two forces W_1, W_2 by merely dropping the weights and measuring their accelerations. For although very dissimilar, their acceleration is the same, the physical constant g. Again the proportionality between force and acceleration is possible; but if so, in the equations

$$W_1 = m_1 g$$

and

$$W_2 = m_2 g$$

m_1 and m_2 are to be inserted as unequal multiplying coefficients. In this particular example

$$\frac{W_1}{W_2} = \frac{m_1}{m_2}$$

and the ratio between the forces equals the ratio between the m coefficients.

4. Newton's account of the principles of motion

To appreciate Mach's critique of Newton's dynamics, it is necessary to give a part of Newton's own statement.[1]

[1] I. Newton, *Mathematical Principles*, pp. 1, 5, 13, 14, 15.

DEFINITION I

The quantity of matter is the measure of the same, arising from its density and bulk conjointly.

. . . It is this quantity that I mean hereafter everywhere under the name of body or mass. And the same is known by the weight of each body, for it is proportional to the weight, as I have found by experiments. . . .

DEFINITION II

The quantity of motion is the measure of the same, arising from velocity and quantity of matter conjointly.

. . . Hence it is, that near the surface of the earth, where the accelerative gravity, or force productive of gravity, in all bodies is the same, the motive gravity or the weight is as the body; but if we should ascend to higher regions, where the accelerative gravity is less, the weight would be equally diminished, and would always be as the product of the body, by the accelerative gravity. . . .

LAW I

Every body continues in its state of rest, or of uniform motion in a right line, unless it is compelled to change that state by forces impressed upon it. . . . .

LAW II

The change of motion is proportional to the motive force impressed; and is made in the direction of the right line in which that force is impressed. . . .

LAW III

To every action there is always opposed an equal reaction: or, the mutual actions of two bodies upon each other are always equal, and directed to contrary parts. . . .

COROLLARY I

A body, acted on by two forces simultaneously, will describe the diagonal of a parallelogram in the same time as it would describe the sides by those forces separately. . . .

COROLLARY II

And hence is explained the composition of any one direct force AD, out of any two oblique forces AC and CD: and, on the contrary, the resolution of any one direct force AD into two oblique forces AC and CD: . . .

To make the meaning of the two corollaries clear, Newton adds two diagrams.

5. Mach's first criticism of Newton's dynamics

Mach is second to none in his admiration of Newton:

> Newton's reiterated and emphatic protestations that he is not concerned with hypotheses as to the causes of phenomena, but has simply to do with the investigation and transformed statement of *actual facts*,—a direction of thought that is distinctly and tersely uttered in his words "hypotheses non fingo", (I do not frame hypotheses)—stamps him as a philosopher of the *highest* rank. He is not desirous to astound and startle, or to impress the imagination by the originality of his ideas: his aim is to know *Nature*.[1]

For Mach, Newton's "hypotheses non fingo" is a terse expression of the first principle of a correct philosophy of physics. Nevertheless, Newton's formulation of the principles of dynamics can possibly be improved.

Mach's suggestions are reviewed here in the order of Newton's own statements. Newton's account of 'mass' is considered first. Whether Newton was or was not successful in *defining* mass makes no difference to the value of the immense step he was taking. He, unlike Galileo, Huygens and Descartes, realised that there are weights, and there are masses, and that these are not the same.

> His sense of the value of the concept of mass places him above his predecessors and contemporaries. It did not occur to Galileo that mass and weight were different things. Huygens, too, in all his considerations, puts weights for masses; . . .[2]

Such a concept as mass was bound to arise once it was found that the same body can have different accelerations of fall, and, so, different weights. Some time before the *Principia* appeared, Richer had shown that the time of oscillation of a pendulum could be increased slightly merely by taking it from Paris to Cayenne.[3] Newton no doubt felt the universal craving for "substance concepts".[4] What was the *substance* of Richer's pendulum, which must be the *same* in Cayenne as it is in Paris?

[1] M., pp. 236–7. [2] M., p. 312. [3] Ibid. [4] M., p. 329.

It was the mass, the body, the quantity of matter. But, according to Mach, Newton's description of mass as

$$\text{density} \times \text{volume}$$

is circular and therefore illogical. The objection is just if by density is meant the quotient

$$\frac{\text{mass}}{\text{volume}};$$

for then mass is defined in terms of density, and density is defined in terms of mass.

> With regard to the concept of "mass", it is to be observed that the formulation of Newton, which defines mass to be the *quantity of matter* of a body as measured by the product of its volume and density, is unfortunate. As we can only define density as the mass of unit of volume, the circle is manifest. Newton felt distinctly that in every body there was inherent a property whereby the amount of its motion was determined and perceived that this must be different from weight. He called it, as we still do, mass; but he did not succeed in correctly stating this understanding.[5]

Mach's view that Newton's failure in this respect is a matter of communication—a matter of terms and language—is borne out by the fine comment following DEFINITION II quoted above. The simple fact that bodies of different weight fall with the same acceleration shows that the weights, which are forces, cannot be measured by the acceleration only. Force must be computed as

$$\text{acceleration} \times \text{some other factor},$$

and for bodies of the same homogeneous material this other factor is proportional to volume. This other factor is not however merely the volume because equal volumes of wood and iron have the same acceleration of fall and also the same volume; they do not have equal weights. Newton's idea of mass is Aristotelian, and as such is still helpful. In Aristotle's metaphysics 'matter' is the 'potential' which can become 'substance' by having the endowment of 'form'. In Newton's dynamics

[5] M., p. 237. Italics as in German text. One word of translation amended.

'mass' is beyond the reach of sense-perception, or it is a 'non-O' concept. Mach, as we shall see below, defines 'mass' dynamically; Newton views 'mass' statically as 'substance'.

It is not entirely clear to me what Newton meant by the term 'law'. What Newton calls LAW I is in fact an experimental generalisation, or a law in the modern sense of the term.[6] But the relationship between Newton's first and second 'laws' is faulty. As the two statements stand, the first is a necessary deduction from the second. If we use the symbols P, m, f for force, mass and acceleration respectively then the second 'law' is to the effect that

$$P = mf, \text{ when } m \text{ is constant.}$$

Suppose $P = 0$ and $m \neq 0$, then

$$f = 0$$

or

$$v \text{ (velocity)} = \text{constant},$$

which is the first law again.[7] Against this formal objection of Mach, it may be suggested that Newton's separation of LAW I from LAW II led naturally to the concept of 'the inertial frame' which will occupy us in the next chapter.[8] Moreover, the special case

$$v = \text{constant}$$

of LAW II is an empirical generalisation, whereas the more general

$$P = mf$$

is not.

Newton's LAW II is not a law at all in the modern sense of the term; Mach therefore replaces it by two definitions.[9]

LAW III is however a 'law' in our sense. *Prima facie* the experiment with the floating magnets is a dynamical illustration of this empirical law. An important idea behind Mach's restatement is to define mass-ratio in terms of the accelerations produced when two masses are subject to the same force. What could be simpler or safer than to arrange for this *same force* to

[6] M., pp. 332–6. [7] M., p. 332.
[8] This comment was suggested by the Athlone Press. [9] M., p. 302.

be *that between the two masses* mutually interacting? This force could in principle be gravitational, but need not be. In Newton's experiment the force is that between magnets, and weight is entirely irrelevant because all motions are horizontal. Mach takes Newton's experiment as a *mode of definition of the concept of mass.*[10]

It would be unwise to assume that Newton himself was not clear in his usage of the terms definition, law and corollary; but —so Mach argues—nothing is to be gained by the retention of such terms when possibly their meaning has changed. The two corollaries quoted are in fact an experimental proposition, *viz.* the parallelogram of forces.[11] A modern statement of Newton's principles requires each item to be correctly designated as 'experimental proposition' (law) or 'definition'; these two are sufficient terms of designation.

Mach's improved version of Newtonian dynamics appeared first in Carl's *Repertorium.*[12] It is given again, entire, as a note at the end of the *Conservation of Energy.*[13] Further improvements were made before 1883. This is how it appears in the *Mechanics:*

(a) *Experimental proposition.* Bodies set opposite each other induce in each other, under certain circumstances to be specified by experimental physics, contrary *accelerations* in the direction of their line of junction. (The principle of inertia is included in this.)

(b) *Definition.* The mass-ratio of any two bodies is the negative inverse ratio of the mutually induced accelerations of those bodies.

(c) *Experimental Proposition.* The mass-ratios of bodies are independent of the character of the physical states (of the bodies) that condition the mutual accelerations produced, be those states electrical, magnetic, or what not; and they remain, moreover, the same, whether they are mediately or immediately arrived at.

(d) *Experimental Proposition.* The accelerations which any number of bodies A, B, C . . . induce in a body K, are independent of each other. (The principle of the parallelogram of forces follows immediately from this.)

(e) *Definition.* Moving force is the product of the mass-value of a body into the acceleration induced in that body.[14]

[10] M., p. 303. [11] M., pp. 302–3.

[12] E. Mach, 'Ueber die Definition der Masse', *Repertorium für physikalische Technik,* Bd. IV, 1868, p. 355. [13] GG., pp. 50–4. [14] M., pp. 303–4.

With respect to this enlightening programme a few further comments may usefully be made. In the first "experimental proposition" (a) there is a difficulty in the notion of "body". Indeed this difficulty is one which led Newton to delay the publication of the law of gravitation by some twenty years. It can be circumvented by confining the discussion to particles or 'mass-points'.[15]

Mach's comment that the first experimental proposition includes the principle of inertia is an example of his rather extreme partiality for economic expression. The point is that an isolated body has no body "set opposite" to it; it experiences therefore no acceleration.

The definition of mass-ratio (b) takes our minds back to the rectory orchard and the moonlit night of 1666 when the apple fell. If the mass of the apple is m_{A}, that of the earth m_{E}, and the two accelerations g_{A} and g_{E} respectively, then

$$m_{\mathrm{A}}g_{\mathrm{A}} + m_{\mathrm{E}}g_{\mathrm{E}} = 0,$$

where g_{A} and g_{E} are vectors. The two vector accelerations to Newton are in 'absolute space'; the error in taking g_{A} as relative to the earth is minute.

This same definition (b) equally takes our minds back to Newton's experiment with the floating magnets. It is in no way essential that the force which enables mass-ratio to be defined should be gravitational in character. The force can be that between magnets or between electrically charged particles. There is only one kind of moving force; this is the force P which imparts to a particle of mass m an acceleration equal to P/m. This is made clear by Mach's second experimental proposition (c), which is an important addition to Newton's scheme. It is important to stress, as Mach does, that this principle (c) is truly experiential. Suppose for example that A, B, C stand for three masses. Suppose further that experience proves that

$$A = B$$

and also that

$$A = C.$$

[15] R. B. Lindsay and H. Margenau, *Foundations of Physics*, New York, 1957, pp. 92–4.

It does not follow at all that

$$B = C,$$

but a long experience teaches us that it is so.[16] We are not here concerned, as Mach puts it, with a *mathematical* question; we are concerned with a *physical* question. The justice of this view is clear as soon as we recollect that each symbol (A, B, C) conceals *a complex set of operations*, perhaps of the floating magnet variety. Suppose a chemist discovers that the elements A and B combine in the ratio a:b by mass; that further, the elements A and C combine in the ratio a:c by mass. It does not follow that the elements B and C must combine in the ratio b:c by mass. In fact, sometimes they do—sometimes they do not.[17] On first consideration, this illustration may seem a trifle far-fetched. This is only because we tend to revert to the Aristotelian idea of mass as quantity of matter; we forget that the mass equality

$$A = B$$

means that an object with mass A determines an acceleration f in another object of mass B, and that at the same time the second object determines in the first an acceleration of $-f$. Mach is quite justified in comparing this elaborate idea with chemical change itself. Mach gives a parallel illustration from yet a third kind of science.[18] Suppose a body A at 16°C is mixed with a body B at 14°C and the temperature comes to 15°C. Suppose further that body C at 16°C mixed with body B at 14° gives the mixture at 15° also. We may, if we choose to do so, interpret the two statements as follows:

A and B have equal capacity for heat,
and, C and B have equal capacity for heat.

We cannot however infer from these two that:

A and C have equal capacity for heat.

To find out if this is the case, we could have A at 16°C, but we should be obliged to have C at 14°C which is new. In fact it is nearly, but not quite, the case. There could hardly be better or

[16] M., p. 268. [17] M., pp. 268–9. [18] W., pp. 190, 191.

simpler examples of the *treacherous difficulty of applying simple mathematical reasoning to physical questions.*

Once we have discovered the existence of a special *acceleration-determining characteristic* of bodies,[19] and once we have resolutely ceased to think of this characteristic as 'quantity of matter', the question of the conservation of mass throughout chemical and physical change is correctly seen as empirical, a question which only experience and experiment can decide. Landolt's extended investigations (1890–1907) were correctly conceived and philosophically necessary; from them it seemed that total mass is conserved through chemical change. Later (Otto Hahn, 1938) a kind of change was discovered in which mass in fact is annihilated. The relation between Hahn and Landolt is a historical parallel with that between Clausius and Carnot. Carnot (1824) and Clausius (1850) both gave the famous reversible cycle. In Carnot's work heat is conserved; in the revised form of Clausius heat is annihilated to an extent in simple proportion to the mechanical work done. Mach's critical genius shows to great advantage here. As early as 1883, he is considering the possibility of the annihilation of mass.[20] His reflections would not have been less valuable had mass never been annihilated; the fact that it has should prove to the least sympathetic physicist that *there is some value in critical and philosophical analysis of the foundations of physics.*

6. A formal analogy

Suppose the symbol f_1^2 stands for the acceleration produced in the mass particle m_1 by the mass particle m_2. Then the equations for three mass particles m_1, m_2, m_3 may be written:

$$m_1 f_1^2 + m_2 f_2^1 = 0,$$
$$m_2 f_2^3 + m_3 f_3^2 = 0$$
$$\text{and} \qquad m_3 f_3^1 + m_1 f_1^3 = 0.$$

The fact that the ratio m_1/m_3 determined from the third equation has the same value when it is determined from the first two equations is an instance of Mach's second "experimental

[19] M., p. 271. [20] Ibid.

proposition" (the proposition which bears the letter c). More generally, a quotient of the type

$$\frac{f_b^a \cdot f_c^b \cdot f_d^c \cdot \cdot \cdot f_q^p}{f_a^b \cdot f_b^c \cdot f_c^d \cdot \cdot \cdot f_p^q}$$

is *shown by experience* to be either $+1$ or -1.

This acceleration quotient condition and the temperature change quotient condition noted at the end of the second chapter of this book, i.e.

$$\frac{\theta_b^a \cdot \theta_c^b \cdot \theta_d^c \ldots \theta_q^p}{\theta_a^b \cdot \theta_b^c \cdot \theta_c^d \ldots \theta_p^q} = \pm 1,$$

present a complete formal analogy.[1] The four analogous concepts may be tabulated:

Dynamics	Thermodynamics
m	X
f	θ

Mass-ratio cannot conveniently be *defined* by the equation

$$\frac{m_1}{m_2} \equiv -\frac{\theta_2 s_2}{\theta_1 s_1}$$

because the specific heat terms ($s_1\ s_2$) require the notion of equal masses of two different substances. There is however no logical objection to an alternative system of physics in which the quantity X is a major 'non-O' concept defined by the relation

$$\frac{X_1}{X_2} \equiv -\frac{\theta_2}{\theta_1}.$$

In the alternative system, the definition of X-ratio is entirely analogous to the effective (Mach) definition of mass-ratio in classical physics as we have it. Indeed the quantities X and m

[1] W., pp. 190–1.

are very similar; their values are proportional to the volume of a single substance, and depend on specific factors for different kinds of substance. In the evolution of X-ratio physics, the measurement of the temperature terms leading to the quantities θ by means of a gas thermometer *working at a very low constant pressure*, would mark an important advance. That the reversible engine thermometer agrees with the perfect gas thermometer would be a somewhat late discovery in X-ratio physics. It is in the main a historical accident that we naturally think of this proposition in the converse sense.

7. A place for causality

When we discover that the pressure p of a given mass of gas is uniquely determined by its absolute temperature T and volume V, according to the equation

$$\frac{pV}{T} = r \text{ (a constant)},$$

we should not, and do not, think of any one of the quantities p, V, T as the *cause* of any other or others. p, V, T are simply *functions* of each other.[1] The law of refraction of light and Galileo's law,

$$s = \tfrac{1}{2}gt^2$$

are further examples. For these laws, so Mach argues, the general conception of *causality* is useless and empty; it may as well be replaced by *functionality*.

It does not however follow that there is no place for causality in the philosophy of physics. As I have explained earlier in the brief paraphrase of the *Prolegomena*, to Kant, cause and effect are that which persists through change. That something must persist is *a priori* (Kant), and this is regulative in the discoveries of physics. The general regulative principle is itself no particular law of the physical world, but rather a norm for laws, or "a mould of laws" (Boutroux). *What* persists is a matter for experience and experiment. Genuine causal regularities are thus both empirical and also *a priori*. That velocity, momentum,

[1] E., p. 278.

mass, energy and heat are conserved throughout certain physical processes are discoveries of great value because the human mind requires, and is prepared to accept, exemplifications of the Kantian concept of the understanding. The laws which correspond to these discoveries, the laws of conservation of velocity, momentum, mass and energy, are therefore distinguishable from the more direct experimental generalisations such as the two quoted above. Since then there are at least two kinds of law, there is something to be said, against Mach, for retaining the term 'cause' to characterise one of these kinds.

Emile Meyerson refers to the Newton–Galileo law or principle of inertia, and favours the Kantian interpretation of this and other conservation principles:

> . . . the principle of inertia cannot be proved *a priori*. . . .
>
> But if the principle of inertia is capable of being considered as an experimental truth, is it really as such that it . . . forces our assent? . . .
>
> . . . one may ask how . . ., in spite of the fact that only a scarcely convincing demonstration was possible at that moment, the principle of inertia could have been so rapidly accepted as the foundation of all mechanics. . . .
>
> . . . motion, having become a state, transforms itself into an entity, a substance—that is, in virtue of the principle of causality, our mind shows the invincible tendency to maintain its identity in time, to conserve it. The body which is displaced is in a "state of motion". What distinguishes this state from others, what constitutes . . . its particular shade, is its velocity. . . .
>
> . . . Is the principle of inertia *a priori* or *a posteriori*? It is neither . . ., because it is both. . . .
>
> . . . *certain essential things persist*. But this is an indefinite formula, for it does not tell us what are the things which persist . . . experience alone can teach us that.[2]

Mach's desire to eliminate *cause* is a symptom of his desire to eschew *explanation* in physics, his determination to make do with *description*. But I suggest that Meyerson's Kantian view of explanation, as interpretation through causality, is worth retaining. Although Meyerson finds Mach's philosophy inadequate, he accepts Mach's account of the law of inertia as

[2] E. Meyerson, *Identity and Reality*, pp. 140–7.

being derived from a thought-experiment used to extend the actual experiments of Galileo. The principle is an experimental generalisation.[3]

8. Further comment on the concepts of mass

In a valuable critical paper on the concept of mass,[1] G. Burniston Brown has in some measure defended Newton as against Mach. He admits that

$$\text{mass} \equiv \text{volume} \times \text{density}$$

is "a definition of the greatest obscurity", but he considers that the towering intellect of Newton was unlikely to perpetrate a grossly illogical circular definition. He suggests therefore that when Newton "spoke of density he meant what we now call relative density", and so Newton's definition of mass is not circular after all. Mass would then appear to have the physical dimensions of volume. The two masses of two equal volumes of gold and tin would be two measures of the given volume, expressed in terms of two different units of volume.

Newton thought he was defining "quantity of matter". The product

$$\text{volume} \times \text{relative density}$$

suggests 'a lump of substance' or 'a quantity of matter'. It is likely that Newton himself would not have considered Mach's reduction of mass-ratio to acceleration-ratio as entirely satisfactory. We can admire Mach for his profound insight into the notion of mass defined dynamically; our admiration is deepened because of the modern discovery of ways to annihilate mass. But we cannot judge Newton's achievements in such terms. Newton was giving an account of something conserved and persistent. I like Bridgman's comment:

[3] Ibid., pp. 9, 10. (Reference to Mach.)

[1] G. Burniston Brown, 'Gravitational and Inertial Mass', *American Journal of Physics*, **28**, no. 5, May 1960, pp. 475–83. Dr Burniston Brown was the author's director of studies during his work for the London Ph.D. degree. His help and encouragement are here gratefully acknowledged.

In just what sense is matter conserved? Certainly not in terms of mass, as we at one time thought. Nevertheless we undeniably have a feeling that there is some sort of conservation property here, and are driven to formulate it badly in terms of a hypothetically constant number of protons and electrons. I have long thought that Newton was groping after some very similar idea when he so far forgot himself as to define mass as quantity of matter, a definition perfectly meaningless to a rigorous and unsympathetic interpretation.[2]

In the same paper, Burniston Brown proceeds to give a clear and illuminating account of the distinction between *inertial mass* (m_i) and *gravitational mass* (m_a) which cannot, according to the principles of P. W. Bridgman, *a priori* be assumed to be identical.[3] Let us suppose that it is possible and convenient to set up an invariable unit of force. Then the 'operations' for m_a and m_i would be respectively:

I. A particle of unit mass, together with and set opposite to another particle of unit mass 1 cm away, mutually exert the gravitational force equal to the invariable force unit;

II. The force measured by the invariable force unit gives the unit mass an acceleration of 1 cm per second each second. The novel feature here is that both mass concepts (the reader will see the point of the plural in the section heading) are defined in terms of an 'invariable force unit', and that 'mass' thus appears as secondary to 'force' as primary. Gauss wished to form the theory of mechanics in accord with the sequence

$$\text{force} \rightarrow \text{mass}.$$

In practice, there is some difficulty in obtaining an invariable force unit, and, in the end, Gauss was obliged to work with a 'primary' mass unit instead. This makes a full discussion rather complicated but does not affect the principle. As Burniston

[2] P. W. Bridgman, *The Logic of Modern Physics*, pp. 110, 111.

[3] Nevertheless, Newton used one kind of mass only, m_i. He discovered that central gravitational force between two mass points is proportional to the product of their two m_i values. If mass is redefined from the gravitation formula, the discovery that m_i/m_a is a true constant confirms a part of Newton's law of gravitation. See H. Dingle, 'Particle and Field Theories of Gravitation,' *British Journal for the Philosophy of Science*, XVIII, no. 69, 1967, pp. 57–64.

Brown indicates, Gauss's instinct "to begin with force" was perfectly sound:

There is another consideration which supports Gauss's original inclination to prefer a unit of force. We are directly conscious of distance, of time, and of force, whereas we have no direct knowledge of mass. In the first three cases there is direct knowledge, together with measures, which are numbers obtained directly by specified measuring operations; in the case of mass there is only a number—measured indirectly—there is no sensual contact.[4]

This is, of course, explicit support for Mach's theory of the metrical concepts, and even for the slight extension of that theory which I have attempted in this essay. 'Force' is an 'O' concept, and of it Dr Burniston Brown affirms that "there is direct knowledge"; 'mass' is absolutely imperceptible or 'non-O', and Burniston Brown affirms that with mass we can have "no sensual contact".

Mach's definition of mass is "a set of operations" for m_1. This is true even if the definition is operated, so to say, by gravitational force.

9. Two metrical concepts: moment of inertia and couple

The third chapter of the *Mechanics* begins with this bold assertion:

The principles of Newton suffice by themselves, without the introduction of any new laws, to explore thoroughly every mechanical phenomenon practically occurring, whether it belongs to statics or to dynamics.[1]

To illustrate this, Mach gives an original proof of the law of the lever, in which Archimedes' lever principle is deduced from more general equations describing the motion of the *unbalanced dynamic lever*.[2] The proof deserves to be better known; I have set it out here in a slightly modified form.

The diagram (Fig. 21) represents three masses m_1, m_2 and M joined by very stiff elastic connections and placed at the

[4] G. Burniston Brown, loc. cit., p. 479.

[1] M., p. 344. [2] M., pp. 351–7.

vertices of a triangle. Compared with m_1 and m_2, M is 'very large'. The stiff triangle is thought of as lying on a perfectly smooth horizontal table. Later in the argument, the triangle degenerates into a rigid linear lever, and the very large mass M becomes the fulcrum. Evidently the fiction cannot be developed in a vertical plane, for the large force Mg is not wanted.

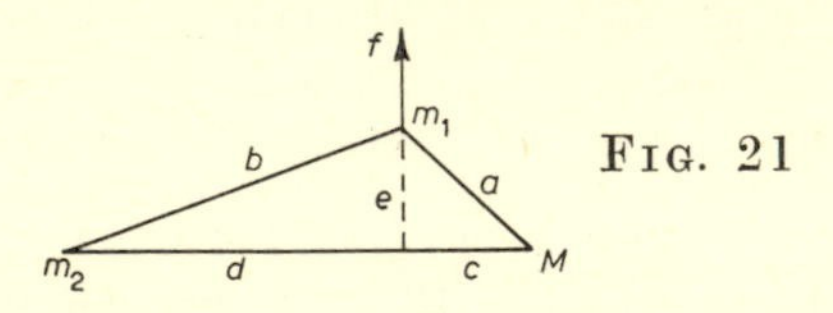

FIG. 21

Mach begins the proof with these words:

> Let m_1, now, receive from the action of some external force an acceleration f perpendicular to the line of junction $Mm_2 = c + d$.[3]

If this statement is correctly understood, the rest of the proof presents little difficulty. The triangular system is not 'absolutely rigid' but, as I have already called it, 'very stiff'. The Newtonian principles cannot be applied to 'rigid' body problems unless the adjective 'rigid' is defined in Mach's own way:

> Bodies in which we purposely regard the mutual displacement of the parts as evanescent, are called *rigid* bodies.[4]

If m_1 receives an acceleration f, this means it is subjected to a force m_1f in accord with Newton's second 'law' (Mach's last definition); the acceleration f is however momentary and is quickly checked by the forces conveyed through the members of the stiff frame. After a short time δt, f is reduced to what is called, somewhat misleadingly, the initial acceleration of the mass m_1 as part of the rigid system.

We suppose that m_2 gives to m_1 a counter acceleration s down along the side b, and M gives it a counter acceleration σ down along the side a. The three accelerations operating on m_1 are shown in Fig. 22. f is quickly reduced by s and σ, the components of which accelerations in the direction opposite to f are

$$\frac{se}{b} \text{ and } \frac{\sigma e}{a} \text{ respectively.}$$

M., p. 352. [4] M., p. 351.

To avoid confusion it must be noted that Mach uses such quantities as f, s and σ as scalar magnitudes. The directions

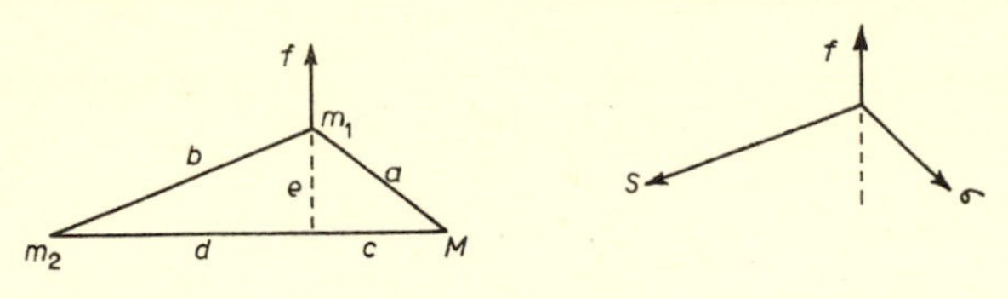

FIG. 22

must be borne in mind as the proof proceeds. Almost immediately after the initial blow the acceleration of m_1 in the f direction (Fig. 21) becomes

$$f - \frac{se}{b} - \frac{\sigma e}{a}. \tag{1}$$

Because of the elastic member b, the mass m_2 acquires an acceleration s_2 (say) up along the side of length b. This acceleration s_2 has components

$$s_2 \frac{d}{b} \text{ parallel to } d \text{ and } c\ (\rightarrow)$$

and

$$s_2 \frac{e}{b} \text{ perpendicular to this } (\uparrow).$$

As a consequence, m_2 approaches M slightly; M can be taken as stationary because its mass is *very large.*

The whole arrangement is too rigid for m_1 to move laterally, i.e. at right angles to the f direction. In other words, s and σ quickly assume values such that

$$s \cdot \frac{d}{b} = \sigma \cdot \frac{c}{a}. \tag{2}$$

This is at least plausible; the elastically held system adjusts itself so that m_1 moves off exclusively in the direction of the initial blow. Newton's third law can now be applied to the equal and opposite forces along the two directions of the side b. Thus:

$$m_1 s = m_2 s_2. \tag{3}$$

In writing equation (3), Newton's second 'law' (Mach's last definition) is employed. From equations (1), (2) and (3), we have acceleration of m_1 equals

$$f - s \cdot \frac{e}{b} - \sigma \cdot \frac{e}{a}$$

$$= f - s_2 \cdot \frac{m_2}{m_1} \cdot \frac{e}{b} - s \cdot \frac{d}{b} \cdot \frac{a}{c} \cdot \frac{e}{a}$$

$$= f - s_2 \cdot \frac{m_2}{m_1} \cdot \frac{e}{b} - s \cdot \frac{d}{b} \cdot \frac{e}{c},$$

and this expression, by a further application of (3), becomes

$$= f - s_2 \cdot \frac{m_2}{m_1} \cdot \frac{e}{b} - \frac{m_2}{m_1} \cdot s_2 \cdot \frac{d}{b} \cdot \frac{e}{c}$$

$$= f - s_2 \cdot \frac{e}{b} \cdot \frac{m_2}{m_1} \left(\frac{c + d}{c}\right)$$

$$= f - \phi \cdot \frac{m_2}{m_1} \left(\frac{c + d}{c}\right). \tag{4}$$

In (4)

$$\phi \equiv s_2 \cdot \frac{e}{b}$$

and is, physically, the acceleration of m_2 in the f direction.

The whole stiff triangular frame is now modified into a linear rigid lever with the masses m_1 and m_2 at distances r_1 and r_2 from the fulcrum M.

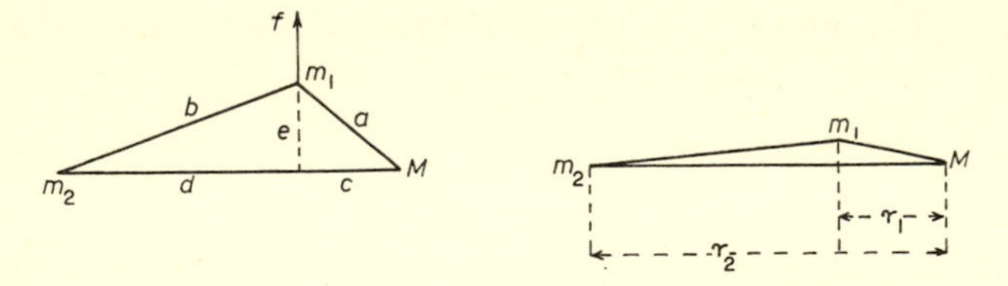

FIG. 23

As shown in Fig. 23

$c + d$ becomes r_2 and c becomes r_1.

During the short initial time δt the accelerations of m_1 and m_2

are variable, but at each instant during δt the equations (1)–(4) hold. The acceleration of m_1 decreases whilst that of m_2 increases so that, at the end of time δt,

$$\frac{\text{acceleration of } m_2}{\text{acceleration of } m_1} = \frac{r_2}{r_1} \tag{5}$$

holds. Then the rigid lever moves as a whole about M. At this stage,

$$\frac{\phi}{f - \phi \cdot \dfrac{m_2}{m_1} \cdot \dfrac{r_2}{r_1}} = \frac{r_2}{r_1} \tag{6}$$

is true, from equation (5). Equation (6) rearranged becomes

$$\phi = \frac{m_1 r_2 r_1 f}{m_1 r_1^2 + m_2 r_2^2}$$

and the corresponding angular acceleration ψ of both m_1 and m_2 is given by

$$\psi = \frac{\phi}{r_2}$$

$$= \frac{m_1 f \cdot r_1}{m_1 r_1^2 + m_2 r_2^2}. \tag{7}$$

This equation shows the applied couple in the numerator and the moment of inertia in the denominator. The two central concepts of rigid dynamics thus emerge out of the analysis which employs only the Newtonian principles. It is instructive to write down the expression for the linear acceleration of m_1, that is

$$\frac{m_1 f r_1^2}{m_1 r_1^2 + m_2 r_2^2}, \tag{8}$$

which is a quantity which must be

$$< f$$

as it ought to be.

Finally let us consider the ordinary Archimedean lever of the first kind. The diagram represents this lever in the vertical plane, the rod connecting the masses being rigid and without mass. Let us suppose that

$$m_1 r_1 > m_2 r_2.$$

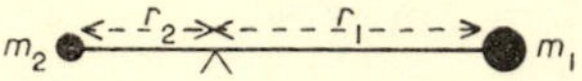

FIG. 24

Then the lever will *begin to accelerate* with the angular acceleration, given according to equation (7) by

$$\psi = \frac{m_1 g r_1 - m_2 g r_2}{m_1 r_1^2 + m_2 r_2^2}. \tag{9}$$

The sense of ψ is clockwise. For $\psi = 0$,

$$m_1 g r_1 = m_2 g r_2$$

which is the ordinary law of the lever. The most celebrated law of statics is thus deduced by putting

$$\psi = 0$$

into an equation giving an angular acceleration, i.e. an equation describing a dynamic state of affairs. The proof illustrates the ideal of Gauss himself, who in 1829 stated that the theorems of statics should emerge out of the more general theorems of dynamics.[5] Rest is simply a special case of motion.

It is worth emphasising what Mach's analysis implies. Suppose the unbalanced system of Fig. 24 is held, and then let go. *Both* m_1 and m_2 *begin* to fall *downwards* with the acceleration g.

The reader may like to work out the whole problem again for the case in which M lies *between* m_1 and m_2.[6]

10. The concepts of mechanical energy

Mach successfully derives the law of the lever, both in its general dynamic form and in the special Archimedes case, from the Newtonian principles. The balanced lever illustrates those

[5] M., p. 441.

[6] J. Bradley, *Ernst Mach's Philosophy of Science*, pp. 603–8.

principles in a much more direct and simple way (Fig. 25). For, according to Newton's third law, the reaction R is a force exerted by the fulcrum on the lever and the numerical value of R is $m_1g + m_2g$. The notion of statical moment of a force (Archimedes and da Vinci) is partial and unsatisfactory. There are not merely two *forces* acting on the lever, but two *couples*;

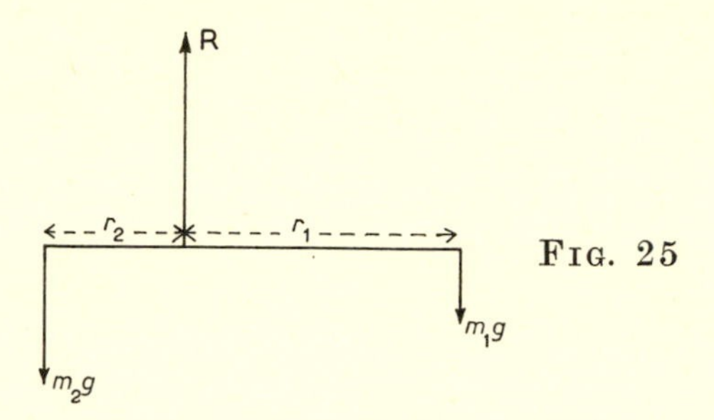

FIG. 25

part of R is taken with m_1g ($\pm m_1g$ at arm r_1) and the other part taken with m_2g ($\pm m_2g$ at arm r_2).

The theory of 'centre of gravity' may be regarded as the law of the lever in reverse. One simply thinks of the system of Fig. 25 in a new way: given the forces m_1g and m_2g, find the fulcrum. This fulcrum is the centre of gravity. This theory in turn can easily be extended to the three dimensions of an *actual* solid rigid body. The notion of 'centre of mass' is a further natural extension.

From the notion of centre of gravity, and Galileo's work on falling bodies and inclined planes, Huygens was able in effect to reach the concepts of potential and kinetic energy. The next diagram (Fig. 26) shows a number of mass points $m_1\ m_2\ m_3\ .\ .\ .$ at vertical distances $h_1\ h_2\ h_3\ .\ .\ .$ above the horizontal mathematical base line. The mass points need not be in the same vertical plane.

Now it is supposed that all the mass-points fall to the base line. The descent of the centre of gravity is given by

$$\frac{\Sigma mg\ .\ h}{\Sigma mg}$$

which is a form of the law of the lever. The mass-points reach the base plane at different times and each with its own speed,

$v_1\ v_2\ v_3$ . . . It is now supposed that each particle rises in a vertical plane along a perfectly smooth incline, or simply that each mass point bounces back to its original height after its

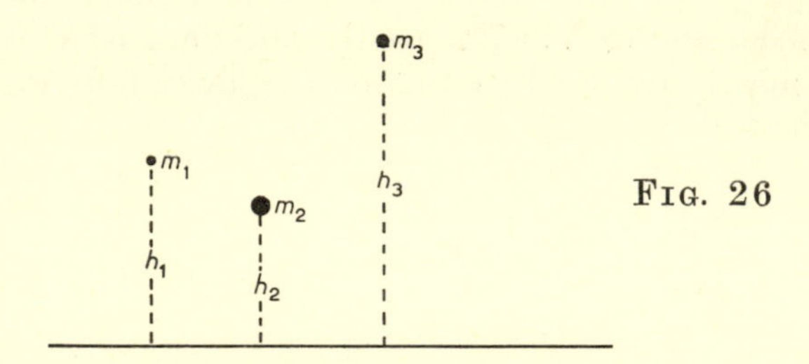

FIG. 26

perfectly elastic collision with the horizontal plane. By Galileo's findings the height attained by the individual mass-point m_1 is

$$\frac{v_1^2}{2g} \text{ (which equals } h_1).$$

The rise of the centre of gravity is therefore

$$\frac{\Sigma mg \,.\, \dfrac{v^2}{2g}}{\Sigma mg}.$$

Huygens simply assumes that *the rise of the centre of gravity must be equal* (*in the ideal case*) *to the fall of the centre of gravity.* In that case

$$\frac{\Sigma mg \,.\, h}{\Sigma mg} = \frac{\Sigma mg \,.\, \dfrac{v^2}{2g}}{\Sigma mg}$$

or,

$$\Sigma mg \,.\, h = \tfrac{1}{2}\Sigma mv^2.$$

These are potential and kinetic energy terms. The last equation becomes in words:

Loss in potential energy of the mass points during their fall	=	Total kinetic energy acquired by them.

Mach gives the same theorem for the solid rigid extended

body.[1] The next diagram (Fig. 27) shows the system under consideration.

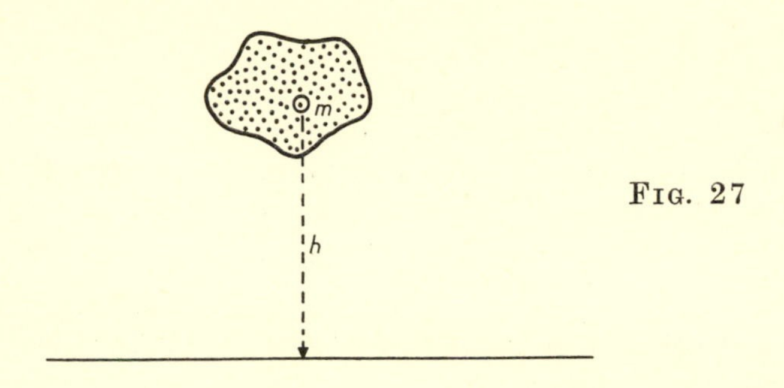

FIG. 27

The shaded object in the diagram represents a solid massive rigid body, made up of a vast concourse of mass elements (m) at distances (h) *vertically* above the mathematical base line. Suppose that the whole rigid body falls to the level of the base, and that it is then dissolved into mass elements, in the manner envisaged by Huygens. One can imagine the act of dissolution occurring over a period, during which each element in turn of the previously rigid body reaches the base line. This is quite in order, because in fact the general proposition proved is true for an array of separate particles, i.e. it is still true, if there is no rigid body at all. By the law of the lever, the fall of the centre of gravity of the rigid body is given by

$$\frac{\Sigma mg \,.\, h}{\Sigma mg}.$$

The various particles of the 'dissolved' body will have different velocities v when they reach the base line. It is supposed that they can now rise in a vertical plane along perfectly smooth inclines. According to Galileo's experiments with the pendulum and the double inclines, each particle will rise to a height determined only by v. The height attained is $v^2/2g$ for the

[1] M., p. 216.

particle whose speed at the base line is v. The rise of the centre of gravity of the separated particles is therefore

$$\frac{\Sigma mg \cdot \frac{v^2}{2g}}{\Sigma mg}.$$

Huygens' principle is slightly extended: *The rise of the centre of gravity of the separate mass elements must equal* (*in the ideal case*) *the fall of the centre of gravity of the rigid body*. So, as before,

$$\Sigma mgh = \tfrac{1}{2}\Sigma mv^2.$$

Again:

Loss in potential energy of rigid body during its fall	=	Total kinetic energy acquired by the 'dissolved' elements of the body.

It is easy to generalise this result. During the motion of any isolated, purely mechanical, system of either particles or rigid bodies or both, the sum 'Potential Energy + Kinetic Energy' is a constant quantity. For this to be true there must be no friction between surfaces, no rolling friction and no friction of bearings.

Huygens himself used this principle in his calculation of the length of the simple pendulum which would oscillate in time with a given compound pendulum.[2] Out of the analysis there emerges a 'moment of inertia' term just as it does from Mach's equations for the dynamic lever. The concepts 'couple' and 'moment of inertia' are necessarily associated with 'rigid body' problems.

The energy concepts are in a different case. One may return to Galileo's experiment (Fig. 15) with the double inclined planes. The ball is idealised and reduced to a mass-point m. Whatever be the inclination of the plane, the speed of the mass-point at the lowest point is given by $v^2 = 2gh$, where h is the common height of all the inclines. It follows that

$$\tfrac{1}{2}mv^2 = mg \cdot h$$

which is *the energy equation, derived for a system with no rigid body and with no plurality of particles.*

[2] M., pp. 210–16.

11. Two more proofs of the general law of the dynamic lever

Although the concepts of mechanical energy need not be thought of as arising exclusively out of the solution of problems in rigid dynamics, their application to such problems has great economic value. Mach illustrates this by several rather complex systems.[1] Nothing of philosophical interest is lost by replacing these complex systems by the simple unbalanced lever depicted in Fig. 24. In the proof which follows, ψ again represents the initial angular acceleration of the system and the other symbols have the same meanings as before.

Consider the small clockwise rotation of the lever which takes place during the very small time δt. The linear acceleration of m_1 is $r_1\psi(\downarrow)$ and that of m_2 is $r_2\psi(\uparrow)$. At end of δt, speed of m_1 will be $r_1\psi\ \delta t(\downarrow)$ and that of m_2, $r_2\psi\ \delta t(\uparrow)$. The total kinetic energy acquired during δt is therefore

$$\tfrac{1}{2}m_1r_1^2\psi^2 \,.\, \delta t^2 + \tfrac{1}{2}m_2r_2^2\psi^2 \,.\, \delta t^2, \tag{1}$$

both terms being positive, and kinetic energy being a scalar quantity. During δt, m_1 falls by $\tfrac{1}{2}r_1\psi\ \delta t^2$ and loses potential energy equal to $\tfrac{1}{2}m_1gr_1\psi\ \delta t^2$. During the same time m_2 gains potential energy equal to $\tfrac{1}{2}m_2gr_2\psi\ \delta t^2$. Hence, total loss of potential energy in δt is

$$\tfrac{1}{2}m_1gr_1\psi\ \delta t^2 - \tfrac{1}{2}m_2gr_2\psi\ \delta t^2. \tag{2}$$

From the fact that kinetic energy gained must equal potential energy lost, it follows that

$$\tfrac{1}{2}m_1r_1^2\psi^2\ \delta t^2 + \tfrac{1}{2}m_2r_2^2\psi^2\ \delta t^2 = \tfrac{1}{2}m_1gr_1\psi\ \delta t^2 - \tfrac{1}{2}m_2gr_2\psi\ \delta t^2,$$

or that

$$\psi = \frac{m_1gr_1 - m_2gr_2}{m_1r_1^2 + m_2r_2^2},$$

as before.

The same formula for ψ (diagram and symbols as before) can be proved by d'Alembert's method (1743). Again, Mach's own

[1] M., pp. 434–40.

examples are more complex.[2] The force $m_1g \downarrow$ on m_1 is partially countered by what d'Alembert[3] called an "equilibrated force" V_1 acting upwards $\uparrow$. V_1 is set up by the action of the rigid rod. The "equilibrated force" V_2 similarly acts upwards $\uparrow$ on m_2 and

$$V_2 > m_2g,$$

so that m_2 begins to accelerate upwards. The various quantities are taken conventionally as positive, and the special case only, as defined above, is considered. Newton's 'second law' gives the two equations

$$m_1\psi r_1 = m_1g - V_1$$

and

$$m_2\psi r_2 = V_2 - m_2g.$$

D'Alembert's principle is simply that V_1 and V_2 are like forces in equilibrium, and that therefore Archimedes' law of the lever, or

$$V_1r_1 = V_2r_2,$$

applies. From these three equations

$$\psi = \frac{m_1gr_1 - m_2gr_2}{m_1r_1^2 + m_2r_2^2}$$

$$= \frac{\text{applied couple}}{\text{moment of inertia}}.$$

One of Mach's reflections on d'Alembert's principle is worth quoting:

> He [Gauss] observed [in 1829], that, in the form which mechanics has historically assumed, dynamics is founded upon statics, (for example, D'Alembert's principle on the principle of virtual displacements,) whereas one would naturally expect that in the highest stage of the science statics would appear as a particular case of dynamics.[4]

Thus, in the Gaussian sense, the d'Alembert proof is logically retrograde; Huygens' solution of the compound pendulum problem is retrograde in the same way.

[2] M., pp. 420–33. [3] M., p. 425. [4] M., pp. 440–1.

12. The conservation of energy-comprehensive

When physicists speak of the conservation of energy they refer to energy of all kinds, including the two main species of mechanical energy and the 'heat substance' of Black. Such a reference is conveniently indicated by the term 'energy-comprehensive'. The two major themes of Mach's physics are the foundations of dynamics and thermodynamics. These are well bound together by Mach's critical commentary.

Before giving this, a note must be added on the term 'work'. At first sight this looks like another name for 'potential energy'. When a mass m is raised against gravity through a height h the force mg does work along its line of action equal to $mg \,.\, h$. This may be regarded either as 'work' done (force $\times$ distance) by the force (mg) or as 'potential energy' (mgh) stored up in (or added to) the 'system'. The separate term 'work' is however needed, because work can be done by a force without a gain of potential energy. When, for example, a match is struck, the work is turned into heat.

Mach gives a full summary[1] of Sadi Carnot's monograph *Sur la puissance motrice du feu*, published in 1824. Carnot knew that heat can do mechanical work and he was inspired in his theoretical analysis of engine cycles by the great and growing use and achievements of the steam engine. He asked the very important question: How much work can heat do? He noticed that it is the *transfer* of heat from a higher to a lower temperature that is essential, not the *consumption* of the heat. Indeed, he was able to retain Black's conception of the conservation of heat, although towards the end of the monograph he expressed some doubt as to whether heat is conserved or not. In 1832, Carnot abandoned the idea of the conservation of heat; Kelvin himself did not do so until 1851.[2]

Carnot found many of Black's ideas serviceable. The temperature difference is at once the essential condition for the production of work, and also the representation of the lack of equilibrium between the levels of heat in the two heat reservoirs. The heat engine can do mechanical work because a change of temperature can bring about a change of volume of the working

[1] W., pp. 215–24. [2] W., p. 282.

material. Conversely, a change in volume of the working material can bring about a change in temperature.

Carnot described in full detail two reversible heat engine cycles, the second of which is the famous one consisting of two adiabatic and two isothermal operations on a permanent gas. When the engine is worked forwards, heat Q, passing from source to sink, does mechanical work, and the tendency is to restore heat equilibrium. On the other hand work can be done on the substance, and the tendency is to produce a lack of equilibrium between the heat reservoirs. This brilliant conception of a reversible cycle, engine and refrigerator, we still use, except for the equality of the two heat quantities associated with the source and sink. Carnot understood quite clearly that the price to be paid for perfect reversibility is in fact a set of ideal operations which can never be quite realised in practice. If there is in fact no temperature difference at all between the high temperature source and gas, as it expands isothermally and does work, no heat can pass from the source to the gas according to the well-known principles regulating the transfer of heat. Of course the difference of temperature may be made as small as one likes. The philosophy of the idea is like that of the differential calculus.[3] An essential condition for reversibility is, that at no stage in the cycle, are two bodies of different temperature in contact. Again, for reversibility, there must be no finite difference between the force of the pressure of the gas in the cylinder and the compression force in the external piston. Carnot's reversible engine cycles are therefore thought-experiments of the same kind as Galileo's experiment with the double inclined planes.[4]

Assuming the truth of what is now known as the second law of thermodynamics, viz. that without interference from outside heat "cannot by itself pass"[5] from a colder to a warmer body, Carnot proved that the work which can be obtained by the transfer of a definite quantity of heat Q between two temperatures t_1 and t_2 cannot possibly exceed that obtained from a perfectly reversible engine, and that this greatest possible amount of work cannot depend on the nature of the working substance in the engine. Guided by the analogy of water falling

[3] W., p. 219. [4] Ibid. [5] Words of Clausius.

from a height and doing work by falling, Carnot surmised correctly that the work done by the reversible engine per cycle is proportional to the temperature difference $t_1 - t_2$.

The present-day student of physics calculates the heat transfers in the Carnot cycle by assuming that they are thermally equivalent to the amounts of work done by the gas in expanding and on the gas in contracting. This from the point of view of Carnot begs the question. Mach's treatment of the subject is, as one comes to expect, a great improvement on this. He notes that it was the work of G. A. Hirn, who like Carnot himself was an engineer, which transformed the subject of thermodynamics into an exact metrical science.[6] For in 1856 Hirn measured as accurately as he could the heat taken from the steam by the steam engine and the heat given up to the condenser, he showed that the latter is significantly less than the former, and that the difference between the two is proportional to the mechanical work done. If one could—I do not think it has ever been attempted—run a Carnot engine between boiling water and melting ice, the heat taken from the source could be judged by the mass of steam condensed, and the heat given to the condenser could be judged by the mass of ice melted; then the heat differential would stand out clearly and be found to be in simple proportion to the work done by the engine. This is a further thought-experiment, but Hirn did something which is sufficiently like it.

The facts may be summarised thus. Heat Q_1 passes from source at t_1, does mechanical work W, heat Q_2 is given up to the sink at t_2, and then, by experiment,

$$Q_2 < Q_1$$

and the ratio $\dfrac{Q_1 - Q_2}{W}$ is constant.

It is possible to retain the conservation of heat substance in spite of these facts; mechanical work then becomes a species of latent heat. This is one of Mach's penetrating insights, and it has already been noticed. On the other hand, if mechanical work is not admitted to be a form of heat, the conservation of heat must be abandoned.

[6] W., p. 241.

13. Mach's theory of temperature (continued from p. 39)

It is now possible to conclude the summary account of Mach's treatment of 'temperature'.

(j) *Kelvin's definition of temperature*

The final outcome of the classical work of Carnot and Hirn is summed up by two equations:

$$Q_1 - Q_2 = cW \tag{1}$$

and

$$Q_1 - Q_2 = d(t_1 - t_2), \tag{2}$$

where c and d are constants. These are the two equations for the perfectly reversible engine cycle. c can be made equal to unity; this is simply the decision to measure work and heat in the same units. Equation (2) was reached by the analogy between a heat engine and a waterfall. It is nearly but not exactly true when the source and sink temperatures are read by any common thermometer. Kelvin's idea was simply to *make* equation (2) exactly true by defining temperature anew. His definition is

$$\frac{T_1 - T_2}{T_1} \equiv \frac{W}{Q_1}, \tag{3}$$

where T is the new concept of temperature. The scope of the advantages arising from the definition (3) can be appreciated by considering the temperatures

$$T_2 \equiv T_1 - 1$$

and

$$T_2' \equiv T_1 - 2,$$

the temperatures T_2 and T_2' differing by 1° on the new scale. Then

$$\frac{T_1 - (T_1 - 1)}{T_1} \equiv \frac{\Delta W}{Q_1}$$

and

$$\frac{T_1 - (T_1 - 2)}{T_1} \equiv \frac{\Delta W'}{Q_1},$$

from which
$$1 = \frac{T_1}{Q_1} . \Delta W$$

and
$$2 = \frac{T_1}{Q_1} . \Delta W',$$

and $\Delta W'$ must be twice ΔW.

Provided the heat-engine-thermometer can be worked from the same source and always takes in the same energy at the source temperature, the temperature differential between the source and sink is proportional strictly to the work done in the cycle and depends on nothing else. No longer need the numerical index of a given 'warmth-condition' depend arbitrarily on the properties of a specific substance.[1]

Kelvin's temperature scale has a natural zero. If in the definition (3) T_2 is made zero, then

$$W = Q_1,$$

and *all* the heat taken in by the working substance from the source at T_1 is transformed into work. So we may affirm that, with the concepts of heat and mechanics defined as they have been, and Mach ultimately takes the view that they might have been defined differently, it is part of the description of thermal phenomena to define temperature in the manner of Kelvin and to discover that *this temperature scale has a natural zero.*[2] But there is not a scrap of evidence that this absolute zero of temperature (Kelvin) corresponds to an absolute cold of warmth-condition.

(k) *Kinetic theory*

Clausius considered that the molecular kinetic theory gives a genuine ground for belief in an absolute cold. When the molecules are at rest, the substance cannot become any colder. Young students of physics may be deceived by this argument, but it is an elementary philosophical error.[3] The kinetic theory

[1] W., pp. 233, 287.

[2] Kelvin also considered another absolute scale of temperature in which $-\infty$ is the 'absolute zero'. See W. Wilson, *Theoretical Physics*, I, London: 1931, p. 295.

[3] W., p. 55, see also p. 341.

is an intellectual construction or play-thing, from which temperature has been excluded *a priori*. There are counters in the game for molecules, counters for their speeds, counters for their elastic collisions and mean free paths, but no counters for temperature. At least one distinguished modern writer agrees with Mach:

> In the kinetic theory of gases, the concept of temperature is primarily a foreign element, since in fact the individual molecules are characterised by their velocity alone.[4]

14. The conservation of energy-comprehensive (continued)

Although it is possible to obtain work from heat, the work W obtained is not quantitatively equivalent to the heat Q, except by means of a thought-experiment with a perfectly reversible engine and a sink at the so-called absolute zero. The order of facts associated with the name of J. P. Joule is, *prima facie*, converse with respect to the order of facts arising from Carnot's cycle. But in Joule's celebrated experiment, in which a falling weight causes water to be stirred and warmed, *all* the potential energy W of the weight system is turned into heat. If Joule's experiment is thought of as a kind of inverted heat engine, in which the object is to turn work into heat, then such an engine is absolutely and readily 100 per cent 'efficient'. On the other hand, the efficiency of the transformation

$$\text{heat} \rightarrow \text{work}$$

cannot rise above the fraction

$$\frac{T_1 - T_2}{T_1}.$$

This asymmetry between the two transformations

$$\text{heat} \rightleftharpoons \text{work}$$

is, and will remain, one of the greatest discoveries in physics. The two transformations are *only roughly the converse* of each other, and the nature of the asymmetry is precisely known and understood.

[4] M. Born, *Atomic Physics*, London: 1952, p. 7.

The experimental discovery of Joule[1] is not strictly that

$$Q = W$$

but that
$$Q = cW,$$

where
$$c = \frac{1}{J}$$

and J is usually called 'the mechanical equivalent of heat'. The idea that heat and mechanical energy are both forms of the same common substance, energy-comprehensive, could not have arisen if c measured by Joule's method did not have the same value as the c of equation (1) in the previous section. Mach makes this simple but important point very clearly:

> . . . only *experience* can inform us . . . whether [these] changes . . . can be made to occur *in the reverse direction.* If such reverse changes did not occur, then the energy principle would lack a sound basis. In this sense consequently the energy principle does rest on experience.[2]

Exactly the same principle has been illustrated earlier in the *Principles of Heat,* in connection with the concept of 'latent heat'. The viability of this concept depends on the fact that latent heat can be transformed back into sensible heat. Since then the quantity c is single-valued we can in fact replace

$$Q = cW$$

by
$$Q = W,$$

just as earlier we replaced

$$Q_1 - Q_2 = cW$$

by
$$Q_1 - Q_2 = W.$$

The conservation of energy-comprehensive appears now simply as an improvement of the mathematical form of these symbolic statements. For suppose C represents the energy-comprehensive of the universe, or of some precisely defined and circumscribed

[1] W., pp. 254–65. [2] W., p. 325.

sub-universe, and that, like Clapeyron in 1843, we take mechanical energy W and heat Q as "quantities of the same kind",[3] then if

$$Q + W = C,$$
$$\Delta Q + \Delta W = 0$$

if C is constant, and the symbol Δ means 'a finite change in'. Joule's experiment is represented by

$$\Delta Q = -\Delta W,$$

the discovery that the *gain* in heat of the water is equivalent to the *loss* in potential energy of the weights. Mach's comment on this is quite just:

Black's "stuff-concept" was inevitably destroyed by Mayer and Joule, and a new, more abstract, more general "*substance-concept*" had to be substituted in its place.[4]

In Joule's experimentally determined values of J ($\equiv 1/c$) there was some 10 per cent variation.[5] This is strong support for Meyerson's view, based on Kant's philosophy, that Joule's mind was *prepared* for a conservation principle, that he *wanted a priori* to regard the two terms of the relationship

$$\Delta Q = -\Delta W$$

as effect and cause respectively, and that he was therefore prepared to ignore the fairly large experimental discrepancy in J values. I think Meyerson is right. There is a place for causality, although Mach will have none of it.[6]

Let us return again to the statement

$$Q + W = C.$$

Such an impressive assertion might well have brought us mental peace at last. But this is not so. No sooner is the grand principle of the conservation of energy formulated, than two more nagging questions make themselves heard. Can mechanical energy and heat be added together in this astonishing way,

[3] W., p. 227 (quotation from Clapeyron). [4] W., p. 319.
[5] E. Meyerson, *Identity and Reality*, pp. 194–5. [6] W., pp. 260, 261.

because 'really' mechanical energy *is* heat? Or is it because 'really' heat *is* mechanical energy? Mach's answer to these subsidiary questions is briefly expressed. The questions have no rational basis, and neither of them need be asked. A very ordinary student of algebra will tell us that from the equation

$$x + y = z,$$

no one can possibly draw a further relationship between x and y.

As a historian of physics, Mach is nevertheless obliged to consider the questions. Let us call them 'the two subsidiary hypotheses'. Some discussion of the first one, viz. that mechanical energy is truly heat, has already been made. Mayer in fact stated in 1867 that "motion is latent heat."[7]

The second subsidiary hypothesis, that heat is mechanical energy, is briefly expressed by the title of the English textbook, *Heat Considered as a Mode of Motion*.[8] That all physics is reducible to mechanics is even more explicitly taught by W. M. Wundt.[9] Mach's comments on these ideas are of great interest. He admits that, for whatever reasons, from the time of Democritus up to A.D. 1896, there has been an undeniable tendency for men to interpret physical events mechanically.[10] The reference to Democritus here is significant because it shows that Mach associates mechanistic philosophy with molecular philosophy. This association is common in modern physical theory; indeed we have improved on the saying of Democritus, that atoms and void alone exist, by the addition of a third term 'motion'. Mach allows himself to reflect on the persistence and apparent vitality of this ancient idea. He notes that although physics is in fact based on all species of sense-perception, 'seen motions' are amongst the most conspicuous and ubiquitous of our experiences. One cannot live long before one finds that a push causes a motion.[11] Not only is this true in the nursery, so to speak, but all the phenomena of physics are in part mechanical. As a bell emits a sound it *vibrates* mechanically, when a bar becomes hotter it *moves* out and becomes longer, there is a

[7] W., p. 250 (quotation from Mayer's *Mechanik der Wärme*, 1867).

[8] J. Tyndall, *Heat Considered as a Mode of Motion*, London: 1865.

[9] W. M. Wundt, *Die Physikalischen Axiome und ihre Beziehungen zum Kausalprincip*, Erlangen: 1866. [10] W., p. 316. [11] W., p. 317.

mechanical force between linear conductors of electricity, and so on. The various phenomena of physics have "a mechanical side"[12] to them. The search for mechanical models of thermal and electrical phenomena is, in part, easy to understand and permissible when exercised moderately. Inevitably the human mind must seek to interpret the less familiar in terms of the more familiar. Indeed this is the simple formula for any kind of educational or enlightening progress. But it is easy to take mechanical models and interpretations "too seriously and too literally."[13]

Yet Mach believed that the thorough-going mechanistic philosophy of his time (1896) would have unfortunate consequences. He believed that atoms and molecules are not 'real in the same way' as the elements of the web of sensuous experience. In my view, he has a very good case. Further, it is simply not true that the class called physical events and the class called 'mechanical events' enjoy the same coverage. The coverage of the former class is much larger. Going right back to Mach's own first principles, 'a feeling of warmth' is just as powerful and pregnant a beginning element as 'a stone seen falling'. In the terms of the present study, all the 'O' concepts have equal rights in the representation of the actual. To assert that temperature is less actual than velocity is nonsense.

15. The convention of the conservation of energy

The expression

$$mgh$$

for the potential energy of a heavy book of mass m, lifted up on to a shelf a height h above the ground, is—and I think should be—extremely puzzling to the unsophisticated mind of the beginner in physics. As we lift the book up we *feel* the force mg, and *feel* or *see* the height h—at least roughly, for force and length are 'O' concepts—, but we have no kind of sense-perception at all of the product mgh, for energy is a 'non-O' concept.

[12] Ibid. [13] Ibid.

When the book is on the shelf it does not look any more energetic than it did on the ground. If this energy

$$mgh$$

is *in* the book, why does the expression for it involve a term h representing the ground which is a distance h *away from* the book? Even more puzzling, why does the expression involve g, which Newton related to the mass and radius of the whole earth? Indeed, the term g has a secondary reference beyond the earth to the whole of the rest of the cosmos. Soon we begin to think that this energy

$$mgh$$

is not 'really' *in* the book. If we replace the book by a helical spring anchored at one end to the ground, and do work on the spring, stretching the other end vertically upwards to the height h, it would be paradoxical to suggest that all the stored potential energy must be in the last ring of the helix, at height h. So we comfort ourselves by thinking of

'the energy of the system',

but we can hardly define the limits of the system.

Perplexed by the scale of such reflections, conjured up by an apparently simple problem in physics, we recollect the facts on which the idea of the conservation of energy is based. After all, there are some facts. For example, let the potential energy *of*(?) the book be reduced by putting a table beneath the shelf and the ground. The potential energy of the book relative to the table top is

$$mgh'$$

where

$$h' < h.$$

If we prefer it, we can, as Joule did, exchange the potential energy of the book for heat, instead of exchanging it for kinetic energy. The heat (again 'non-O' and outside our sensuous apprehension) is calculated by Black's formula. The heat so found at table level, is less than the heat so found at ground level. Potential energy can be *replaced by*, if not exactly *turned into*, heat; just as potential energy can be *replaced by*, if not exactly *turned into* kinetic energy. Moreover we can recover *a definite part* of the heat as mechanical work, and the rate of exchange index is more or less the same for the forward and reverse changes. Slight discrepancies can be eliminated by changing the definition of temperature, or in other ways.

Although the energy principle rests on experience, it is not logically necessary. Mach was one of the first to stress the formal and conventional character of the principle of conservation of energy. He was followed by others. Some seven years after Mach's *Principles of Heat*, Poincaré penned a most enlightening and entertaining account of the same subject. Here is a sample:

> A certain quantity . . . must remain constant. . . . We must take account . . . of the other forms of energy—heat, chemical energy, electrical energy, etc. The principle . . . must be written
>
> $$T + U + Q = \text{a constant},$$
>
> where T is the sensible kinetic energy, U the potential energy of position, depending only on the position of the bodies, Q the internal molecular energy under the thermal, chemical or electrical form. This would be all right if the three terms were absolutely distinct; . . . But this is not the case. Let us consider electrified bodies. The electrostatic energy due to their mutual action will evidently depend on their charge—i.e., on their state; but it will equally depend on their position. If these bodies are in motion, they will act electrodynamically on one another, and the electrodynamic energy will depend not only on their state and their position but on their velocities. We have therefore no means of making the selection of the terms which should form part of T, and U, and Q, and of separating the three parts of the energy. . . .
>
> Of the principle of conservation of energy there is nothing left then but an enunciation:
> *There is something which remains constant* . . . the principle of the conservation of energy being based on experiment, can no longer be invalidated by it.[1]

Mach, and Poincaré in less measure, hope to refrain from metaphysics. A great science like physics requires however a background scheme of metaphysical interpretation. We may

[1] H. Poincaré, *Science and Hypothesis*, pp. 124–8.

find in the concept of energy another exemplification of one of Kant's *Verstandesbegriffe*;[2] our minds require energy to be conserved for the intelligible interpretation of change.

16. The *Ding an sich*

Although both mass and energy-comprehensive may be interpreted in the light of the cause/effect category of Kant, there is a glaringly obvious distinction between the two physical concepts. This is simply that mass enjoys spatial location whereas energy is a calculated attribute of a mechanical or other physical system. The mass of a book on a high shelf is unequivocally *where* the book is; the potential energy of the book not so. The mass of a rigid body has both positive and negative analogies with the *Ding an sich* of Kant; both have location in space and both are absolutely beyond the possible reach of sense-perception.

Any 'non-O' concept, and the concept of mass in particular, might well have led Mach to revise his idea of the 'common thing' being *merely* a 'family of sense-data'. I may look at the sun, see its disc and feel its directed warmth. I may refer to the 'family' of sense-perceptions which are localised as the sun, and even consider this family as located in space 93 000 000 miles away. Even taking into account the possible sense-perceptions which I may have at other times and which other people may have, it is a violation of common sense and a grossly fallacious extrapolation of actual intermittent experience to press the permanent and continuously persistent gravitational forces between the sun and this earth and the other planets into the same family. But these forces are of the greatest physical significance in classical physical mechanics. It is desirable to suppose therefore that in addition to the family, there is also a "physical occupant" or ordinary physical object in the same place as the family; this ordinary physical occupant is affected by gravitational forces, and it could even be identified with Kant's *Ding an sich*.

As this is perhaps the most important flaw in Mach's general

[2] E. Meyerson, *Identity and Reality*, pp. 189–214.

philosophical position, the argument deserves reiteration in the more expert words of Professor Price:

Thus for thousands of years a rock continues without any intermission to resist the pressure of other rocks piled up above it. And it seems extremely likely that so long as a physical occupant continues to exist and to occupy space it is *always* actually manifesting some causal characteristics or other. Most of them, no doubt, suffer intermissions on occasion. . . . But probably there are some few which continue to be actually manifested so long as the physical occupant continues in being at all; e.g. gravitational attractiveness, . . . If we insist on attributing causal characteristics to families (or to their visuo-tactual solids) we must radically alter our conception of what a family is. We must say that it has a *further constituent* over and above the actual and obtainable sense-data—a constituent of an entirely new kind, whose mode of situation in space and of persistence through time are both independent of, and different in their very nature from, the mode of situation and of persistence which belongs to the *sensory* part of it. We shall then have to admit that the family is more than a mere system of sense-data, . . . And it seems better to mean by 'family' . . . simply a system of actual and obtainable sense-data, . . .; and to call the causally-characterized something, the physical occupant, by its accustomed name of physical object. And it is quite clear that . . . the family is *not* the physical occupant . . .

We must conclude, then, that Phenomenalism is false. . . .

So far, Kant's conception of the *thing in itself* seems to be substantially justified.[1]

Briefly, it is difficult to see how science can be *exclusively* about families of sense-data, if it is admitted that science is also about 'non-O' concepts.

[1] H. H. Price, *Perception*, pp. 291–6.

6 The Mach Principle

1. The relativity of physical knowledge

There is no Mach Principle in Mach's writings. His general point of view leads him to recur to the evident relativity of measurement; if I say that the coffee table is 3 feet in length, I can only mean that the foot ruler will fit into the length of the table three times. As physical science is based on measurement —although it is not true that 'science is measurement'—our physical knowledge must be in some degree 'relative', for the basis of it is acknowledged to be 'relative'.

In the simple example of the length of the coffee table, another wooden object, the foot ruler, 'enters' as it were into the measured length. According to Mach, the mass of a body is conditioned in exactly the same way, but with additional complications. The mass of the body can only be found by letting it react with *another* mass, observing accelerations and then making a calculation. This leads Mach to the idea of mass as a relation; he belittles the notion of the *localised* mass, that "obscure mysterious lump" which "we seek for in vain outside the mind".[1] Nevertheless, the notion of the localised mass is of some importance as I have indicated in the previous chapter.

Physical time has the same kind of ineradicable relativity. What does g mean? Simply that when the earth has performed 1/864 000 part of its revolution, a body has fallen from rest through a height of $1/2\ g$ metres towards the earth's centre.

Instead of referring events to the earth we may refer them to a clock, . . . Now, because all are connected, and each may be made the measure of the rest, the illusion easily arises that time has significance independently of all.[2]

[1] P., p. 203. [2] P., pp. 204, 205.

One wonders how seriously Mach wrote those words: "because all are connected". The clock pendulum and the spin of the earth bear what could be called a *metrical relationship* to each other. It does not seem to follow that they must bear an *influential relationship* to each other. Quite simply, it hardly seems necessary to assert that the spin of the earth exerts a force or some other kind of influence on the clock. One is led to reflect that Mach is tacitly introducing ideas from his general philosophical monism. If the *seen* pendulum bob and the *observed* terrestrial spin are elements within the one 'web' or 'porridge' which is the unity of all the elements of experience, then probably any individual element must have some kind of 'pull' on every other. Moreover the term 'pull' could mean or at least include a physical force.

As a fourth example the concept of temperature may be cited. Suppose we read a thermometer and discover the temperature of some water is 15°C. This means that the temperature lies 15/100 of the way up an even scale erected between the temperatures of melting ice and boiling water at a standard pressure. It is not so much assumed that these temperatures are fixed; rather the question of their constancy is not raised. As Mach points out, we experience "a *great relative stability* of our environment",[3] without which science would be impossible. Although, as again Mach indicates in his detailed discussion of the concept of temperature, there is no 'real' temperature lying behind 'measured' temperature, yet we do seem to know that the 'temperature' of melting ice does not alter much. The term 'temperature' in this last statement cannot refer to metrical temperature. *Temperatur* is metrical and must be relative; *Wärmeempfindung* is not metrical and could be absolute. Mach, wisely enough, does not press his account quite so far, but I think this is the logical terminus of his line of thought.

Mach admired J. B. Stallo's *Concepts and Theories of Modern Physics* which came out in 1882, one year before Mach's *Mechanics*. It appears that Mach did not read Stallo's book until considerably later. The following extract could be regarded as the classical statement of the relativity of physical knowledge:

[3] P., p. 206.

Objects are known only through their relations to other objects. They have, and can have, no properties, and their concepts can include no attributes, save these relations, or rather, our mental representations of them. Indeed, an object can not be known or conceived otherwise than as a complex of such relations. In mathematical phrase: things and their properties are known only as functions of other things and properties. In this sense, also, relativity is a necessary predicate of all objects of cognition.[4]

This may be compared with Mach's own words in *Knowledge and Error*, a passage which approximates to what we now call a Mach Principle, yet at the same time is not based exclusively on the facts of dynamics:

Nature is not at all a juggler out to humbug us, nevertheless natural events are combined in an extreme degree. Beyond the circumstances whose mode of connection we wish to investigate in a given case, and towards which our attention is directed for the time, there lies a mass of other circumstances which jointly determine the events, and which conceal the connection which interests us, which complicate and apparently *falsify* the event which is under inspection. Therefore is the investigator not to leave out of account intentionally any *neighbouring circumstance* which may play a part, and he must take into consideration all the *sources of error*.[5]

2. Newton's spinning bucket

Newton postulated absolute space and absolute time, although he realised that there may be no body at rest in the absolute space. Also he was aware of the fact that no reference can be made to a fixed position in the absolute space, which is featureless as well as infinite. Nevertheless he considered that absolute motion could be discerned in a specific kind of dynamical experiment, viz. an experiment which involves the rotation of a fluid. In such an experiment, absolute and relative motion could be distinguished. The relevant passage from the *Principia* runs as follows:

For it may be that there is no body really at rest, to which the places and motions of others can be referred. . . .

[4] J. B. Stallo, *The Concepts and Theories of Modern Physics*, London: 1900, p. 134. [5] E., p. 123.

> The effects by which absolute and relative motions are distinguished from one another, are centrifugal forces, or those forces in circular motion which produce a tendency of recession from the axis. For in a circular motion which is purely relative no such forces exist; but in a true and absolute circular motion they do exist, and are greater or less according to the quantity of the [absolute] motion.[1]

There follows an account of the spinning bucket experiment, which Newton carried out for himself. In demonstrating the experiment to students, I use a toy bucket two-thirds filled with water and suspended by about 3 ft of twisted string. The facts are as Newton states. At first, when the bucket begins to spin, the water 'does not get going'. According to Newton's interpretation its motion in absolute space remains unchanged and the surface of the water remains horizontal.[2] Later, because of the viscous force between the walls of the bucket and the water, the water begins to spin in absolute space, and the effect is that the surface of the water becomes concave upwards. Newton's analysis of the facts is characteristically clear. It is when the motion of the water relative to the walls of the bucket is greatest that the surface of the water remains horizontal; it is when the motion of the water relative to the sides of the bucket has become small or zero that the water creeps up the sides of the bucket. In this way, Newton thought he could demonstrate both absolute and relative spin of the water, and distinguish clearly between them. Moreover Newton anticipated what is now called the Mach Principle by an explicit reference to the fixed stars:

> It is indeed a matter of great difficulty to discover, and effectually to distinguish, the true motions of particular bodies from the apparent; because the parts of that immovable space, in which those motions are performed, do by no means come under the observation of our senses. Yet the thing is not altogether desperate; . . . For instance, if two globes, kept at a given distance one from the other by means of a cord that connects them, were revolved about their common centre of gravity, we might, from the tension of the cord,

[1] I. Newton, *Mathematical Principles*, pp. 8, 10. T. McCormack's translation used, see M., p. 277.

[2] This sentence was recast in its present form after consultation with Professor Dingle.

discover the endeavour of the globes to recede from the axis of their motion, . . . And thus we might find both the quantity and the determination of this circular motion, even in an immense vacuum, where there was nothing external or sensible with which the globes could be compared. But now, if in that space some remote bodies were placed that kept always in a given position one to another, as the fixed stars do in our regions, we could not indeed determine from the relative translation of the globes among those bodies, whether the motion did belong to the globes or to the bodies. But if we observed the cord, and found that its tension was that very tension which the motions of the globes required, we might conclude the motion to be in the globes, and the bodies to be at rest; . . .[3]

3. Berkeley and the relativity of motion

In 1721 Berkeley wrote a tractate called *De Motu.* It is a criticism of Newton's mechanics directed against the notions of absolute space and absolute motion. There is a remarkable resemblance between what Berkeley says and Mach's discussion of the same subject in the *Mechanics.* The fact that Einstein was inspired by Mach's *Mechanics* rather than by Berkeley's *De Motu* is no more than accidental. Neither Mach nor Einstein had read *De Motu.*

It is absurd, argues Berkeley, to imagine that motion is impossible as Zeno did. Motion is a natural evident simple given thing. These comments could have been written by Mach or Bergson:

. . . motion is clearly perceived by the senses, . . . As for the things perceived by the senses, they can scarcely be made clearer or better known by definition.[1]

But the motion which is perceived is relative and the idea of an absolute space is useless:

We tend to be deceived because when in imagination we have abolished all other bodies, each of us nevertheless assumes that his own body remains . . . no motion can be understood without some

[3] Ibid., p. 12. cf. M., pp. 278, 279. Motte's translation; only part of this passage is quoted by Mach.

[1] G. Berkeley, *Philosophical Writings*, edited T. E. Jessop, London: 1952, pp. 209, 210.

determination or direction, which cannot itself be understood unless, besides the body moved, we suppose our own body, or some other, to exist at the same time. . . . Since, then, absolute space never appears in any guise to the senses, it follows that it is utterly useless for distinguishing motions.[2]

Berkeley goes further and, no doubt with the *Principia* in mind, suggests simply that motion and rest relative to the "heaven of the fixed stars" could *replace* the useless absolute motion and rest:

. . . for the determination of true motion and true rest in a way that would remove ambiguity and serve the purpose of mechanical philosophers, who view the system of things more broadly, it would be enough to suppose, instead of an absolute space, relative space bounded by the heaven of the fixed stars, this being regarded as at rest.[3]

And all this is simply another example of the first principle of Berkeley's metaphysics, that *a thing is as it is perceived*:

In physics, sensation and experience, which have to do with manifest effects only, have their place. . . . The physicist considers the series or successions of sensible things, . . .[4]

This is a remarkable adumbration of Mach's views, and the views of positivists in general.[5]

Berkeley had in fact already touched on Newton's doctrine of absolute motion in his earlier work, the *Principles*. In the passage now quoted, Berkeley indicates that an adequate 'relativity physics' is not *exclusively* concerned with *seen* relative motions:

But tho' in every motion, it be necessary to conceive more bodies than one, yet it may be that one only is moved, namely that on which the force causing the change, in the distance or situation of the bodies, is impressed. For however some may define relative motion, so as to term that body *mov'd*, which changes its distance from some other body, whether the force causing that change were impressed on it, or no: yet I can't assent to this, for since we are told, relative motion is that which is perceiv'd by sense, and regarded

[2] Ibid., pp. 212, 213. [3] Ibid., p. 213. [4] Ibid., p. 215.

[5] K. Pearson, *The Grammar of Science*, Everyman's Library, London: 1937, pp. 118, 119.

in the ordinary affairs of life, it follows that every man of common sense knows what it is, as well as the best philosopher: now I ask any one, whether in his sense of motion, as he walks along the streets, the stones he passes over may be said to *move*, because they change distance with his feet? To me it appears, that tho' motion includes a relation of one thing to another, yet it is not necessary, that each term of the relation be denominated from it.[6]

The man walking along the street has a good reason for preferring the hypothesis that it is he who is moving, and not the stones. This is no adverse criticism of Newton. Indeed it is simply a characteristically homely reduction by Berkeley of Newton's own argument as quoted in the previous section. The man who set the two globes spinning or who *felt* the force in the connecting cord would reject the hypothesis of the spinning stellar background. The late Sir Arthur S. Eddington made the same point. Suppose a train accelerates from rest and moves off from its position of rest at the platform. Then there may be nothing to stop us taking the train as a non-inertial frame of reference, relative to which the platform and city are accelerating. But the driver of the train has a solid reason for preferring the more prosaic account of the event. *He knows he has powered the train.*

The complication arises because a *relative acceleration* can be produced by an *absolute force.* There is a relative acceleration of a train and a station platform which (because it is relative) should not be attributed to one rather than the other. But the force causing it is applied to the train. . . . There is a very natural tendency to locate the acceleration in the body which experiences the cause of the acceleration, and say that the train experiences an absolute acceleration and the station no absolute acceleration.[7]

As the matter may be of some importance, a more formal statement is attempted. Suppose two bodies A and B of exactly equal status move about in any way. Then if v is the velocity of A relative to B at any time, then $-v$ is the velocity of B relative to A at the same time. But suppose now that body A is 'powered from within' and that body B is not. A may be a

[6] G. Berkeley, *The Principles of Human Knowledge,* p. 96.

[7] *Isaac Newton,* The Mathematical Association, Memorial Volume, edited W. J. Greenstreet, London: 1927, p. 1.

locomotive burning oil, or a human being burning biscuits. B may merely be a signal post on the line, or the stones in the street. Then it is undeniable that v and $-v$ are the relative velocities as before. But the fact that A is 'powered' and B is not adds an element of asymmetry to the question. The man in the engine, or the mind behind the biscuits, might well argue that *A's velocity v has a definable qualitative difference from B's velocity —v*. For convenience of reference this will be called here the Newton–Berkeley–Eddington principle.

In the *Principles*, Berkeley himself noticed that at the outset of Newton's spinning bucket experiment the water "has, I think, no motion at all"[8] for, so far, it has not suffered from an impressed force. Quite correctly in my view, Berkeley thus applied the Newton–Berkeley–Eddington principle to the great experiment which was to inaugurate relativity physics.

4. Mach's discussion of Newton's bucket

Teachers of science to children know that it is almost impossible to obtain a candid unelaborated statement from a child about *what simply has happened*. What, inquires Mach, is the correct report of Newton's spinning bucket experiment?

I. When the water has a large spin relative to the sides of the bucket, and little or no spin relative to the walls of the room or to the fixed stars, the surface of the water remains horizontal.

II. When the water has little or no spin relative to the sides of the bucket, and a marked spin relative to the walls of the room or to the fixed stars, the surface of the water becomes concave upwards. No amount of wishful thinking, not even by Newton, can prevent the experiment from being what it most certainly is, *a comparison between two relative motions*.

> By absolute rotation he [Newton] understood a rotation relative to the fixed stars, and here centrifugal forces can always be found. . . . The resting sphere of fixed stars seems to have made a certain impression on Newton as well. The natural system of reference is for him that which has any uniform motion or translation without rotation (relatively to the sphere of fixed stars).[1]

[8] G. Berkeley, *The Principles*, p. 98.
[1] M., pp. 280, 281.

Mach's criticism so far depends on a *metrical reference* to the fixed stars, not on some supposed *influence* or *force* between the stars and the bucket of water.

But he goes further than this. "Try", so he invites the reader, "to fix Newton's bucket and rotate the heaven of fixed stars and then prove the absence of centrifugal forces."[2] Even Mach lacks the temerity to predict explicitly what would happen if we were able to carry out this cosmic experiment. Yet we are intended to speculate that if the same relative motions were set up, from the side of the fixed stars instead of from the side of the bucket, then the concave water surface would again be formed. If such speculation is not resisted, there is the further implication that in this thought-experiment there must be an *influence* or *force* between the fixed stars and the water in the bucket. Mach's words, "then prove the absence of centrifugal forces", are intended as a jest; for if the centrifugal forces failed to appear, this might be interpreted as showing that Newton's proof of absolute motion is correct after all. Mach believes that such a proof is essentially impossible. It is important to notice that Mach leaves this comment to the reader, although it can hardly be doubted that such thoughts passed through his mind. The lesson he would have the reader draw is more modest: *Give a candid account of the actual events which take place in the experiment as you have been able to perform it. You cannot whirl the stars round the bucket, and neither you nor I knows what would happen if you could.*

Mach's further criticism of the bucket experiment is similar in quality.

The universe is not *twice* given, with an earth at rest and an earth in motion; but only *once*, with its *relative* motions, alone determinable. . . . Newton's experiment with the rotating vessel of water simply informs us, that the relative rotation of the water with respect to the sides of the vessel produces *no* noticeable centrifugal forces, but that such forces *are* produced by its relative rotation with respect to the mass of the earth and the other celestial bodies. No one is competent to say how the experiment would turn out if the sides of the vessel increased in thickness and mass till they were ultimately several leagues thick. The one experiment only lies before

[2] M., p. 279.

us, and our business is, to bring it into accord with the other facts known to us, and not with the arbitrary fictions of our imagination.[3]

Much the same comments can be made again. Inevitably we are led by Mach's words to consider that the growing walls of the bucket may lead to a bucket whose mass is appreciable relative to the mass of the whole cosmos; that this combination of large mass and small distance from the water may have some kind of *influence* or *force* on the water. Indeed, we may even find ourselves thinking of mass m in the numerator and distance r in the denominator of Newton's formula for gravitational force between masses. After such speculation, Mach's actual words are sobering: "No one is competent to say how the experiment would turn out. . . ."

Mach is cautious. Nevertheless his thought is here ranging beyond that of Newton and Berkeley. Newton considered that his two globes tied by a cord could spin in an immense vacuum; the advantage of having the stars, so to say, is that the observer can correctly assign the motion to the globes. It did not, I think, occur even to Newton that the presence or absence of the stars could affect the tension in the cord between the globes. In C. Neumann's form of the bucket problem, Neumann is content merely to repeat Newton's own doctrine. Mach is prepared to go beyond this doctrine. Neumann supposed that a heavenly body be thought of as rotating about its axis, and that because of this rotation it has become oblate. Suppose further that now *all the rest of the universe is annihilated*, so that the rotating heavenly body is the only body left in the entire cosmos. Then, according to Neumann, the rotation of this isolated body will continue, and the oblate shape will persist. This proves that the rotation must be absolute. If it were relative, it would not be possible to distinguish between the isolated body in rotation and the isolated body at rest.[4] The argument is, as Mach says, seductive (bestechendst). But, he comments:

When experimenting in thought, it is permissible to modify *unimportant* circumstances in order to bring out new features in a given

[3] M., p. 284. [4] M., p. 340.

case; but it is not to be antecedently assumed that the universe is without influence on the phenomenon here in question.[5]

Neumann's problem leads Mach to admit that the universe, or stellar background, could conceivably have some influence on a material body. The admission is gently expressed as a double negative: it is *not* to be antecedently assumed that the stellar background is *without* influence on some material body. In other words, the supreme 'control experiment', that in which the investigator carries out his experiment first in the presence of the rest of the universe and then in the absence of the rest of the universe, cannot be done. Mach is certainly correct to this extent; it is idle for Neumann or anyone else to imagine otherwise. I venture to add one further comment on Neumann's 'experiment'. The cosmic giant who sets the heavenly body spinning would know, by his fatigue, that he has done so. Naturally, he does not expect the sudden annihilation of the rest of the universe to cancel his labour. This is perhaps why Neumann's argument is "seductive". Again the Newton–Berkeley–Eddington principle may apply. The point is missed by Mach.

It is of great interest to review the law of inertia, the law of gravitation and the definition of mass in the light of these ideas on relative motion. Mach himself has made it clear that the first of these is correctly termed a law or experimental generalisation. It is worth inquiring into the character of the motions to which the law refers. As the experiments were done by Galileo, evidently the motions described are relative to the walls of a room somewhere in Italy, probably in Padua or Florence. A law of physics cannot contain a reference to a room in Northern Italy; nevertheless to generalise from the walls of Galileo's room to absolute space, as Newton did, is to get ourselves into fairyland. It would certainly be more correct and modest to replace Galileo's room by the stellar background, as Mach suggests:

The comportment of terrestrial bodies with respect to the earth is reducible to the comportment of the earth with respect to the remote heavenly bodies. If we were to assert that we knew more of

[5] M., p. 341.

moving objects than this their last-mentioned, experimentally-given comportment with respect to the celestial bodies, we should render ourselves culpable of a falsity. When, accordingly, we say, that a body preserves unchanged its direction and velocity *in space*, our assertion is nothing more or less than an abbreviated reference to *the entire universe.*[6]

Newton's first law of motion does indeed contain a *reference* to the entire universe; but this reference need not mean that the entire universe *influences* or constrains a body which is moving in accordance with the first law. Mach's comment, which appeared in the first edition of the *Mechanics*, has attained some notoriety:

The most important result of our reflections is, however, *that precisely the apparently simplest mechanical principles are of a very complicated character, that these principles are founded on uncompleted experiences, nay on experiences that never can be fully completed, that practically, indeed, they are sufficiently secured, in view of the tolerable stability of our environment, to serve as the foundation of mathematical deduction, but that they can by no means themselves be regarded as mathematically established truths but only as principles that not only admit of constant control by experience but actually require it.*[7]

The same kind of critical evaluation must be made of the Newtonian law of gravitation.[8] Fig. 28 shows two mass points

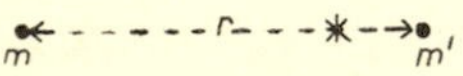

FIG. 28

m and m' separated by a distance r. It is true that, according to Newton's law of gravitation, the masses will generate, in each other, *and relative to each other*, equal and opposite accelerations, the value of each relative acceleration being numerically

$$G\frac{m+m'}{r^2}.$$

But this does no sort of justice to the law of gravitation. What Newton asserted was that the acceleration of m is equal to

$$G\frac{m'}{r^2},$$

[6] M., pp. 285, 286. [7] M., pp. 289, 290. [8] M., p. 282.

and that the acceleration of m' is

$$-G\frac{m}{r^2}.$$

These quantities cannot be computed in absolute space, so that a third body, represented in Fig. 28 by the star in the line mm', is needed to give meaning to Newton's law of gravitation. If one wishes to generalise the example represented in Fig. 28, four stars are sufficient to determine a three-dimensional space for the interplay of measured gravitational forces.

Mach's mode of definition of mass gives further point to this example. Suppose (Fig. 28) that the accelerations set up in m and m' are respectively f and f'. Then, according to Mach,

$$\frac{m}{m'} \equiv -\frac{f'}{f}.$$

If f and f' were relative accelerations, i.e. of m relative to m' and of m' relative to m, then Mach's definition would pronounce that all masses are equal, which would be absurd. If one could, as a pure mind, look at m in complete isolation, and then look at m' also in complete isolation, one would know nothing at all about their accelerations or, indeed, whether they were accelerated at all. If one could, again as a pure body-less mind, look at m as relative to m', and at m' as relative to m, then both particles would have equal and opposite accelerations in all circumstances. Evidently some other body or stellar background is needed; without it one cannot discover or state the law of gravitation, and mass cannot be defined in Mach's special way. The definition of the mass ratio m/m' cannot be achieved by the use of the centre of mass of m and m' instead of the star in Fig. 28; for the position of the centre of mass is determined by the ratio m/m'.

5. The Mach Principle (Mach)

Einstein and other more recent writers have honoured Mach by attributing to him a principle which cannot fairly be found in his writings. It is necessary therefore to draw a distinction between the Mach Principle (Mach) and the Mach Principle

(Einstein). The following statement of the former is based, item by item, on Mach's books.

There is no such thing as an isolated or detached element of experience. All that has been our experience, all that can be our experience and all that could conceivably be our experience, is One. This conclusion is reached through the formulation of a philosophy of the elements of experience; it is also reached through a critical analysis of Newton's dynamical principles, and of the formulation of Newton's law of gravitation. The proper task of science is the direct or indirect description of the One. The description of a 'simple' event, like the motion of a ball, requires a reference to the natural background which is not the ball itself. This is most clearly seen to be true with reference to the metrical attributes of the ball's motion. There is therefore a *metrical* connection between *any individual thing* and *the rest of the universe*, or the rest of the One. The possibility that, beyond this, there is an *influential connection* between any individual thing and the rest of the universe need not, and ought not, to be ruled out *a priori*.

6. Special relativity

When Mach died in 1916, Einstein wrote a glowing obituary appreciation in the *Physikalische Zeitschrift* of 1 April.[1] In the course of this notice, both the special and general theories of relativity are traced back to the influence of Mach.

Professor Dingle begins his book on the special theory of relativity with these words: "The Principle of Relativity may be stated thus: *There is no meaning in absolute motion.*"[2] Now Mach undoubtedly stated this principle over and over again. But Berkeley had already stated it more than a century earlier.

Philipp Frank and P. W. Bridgman have indicated precisely how it is that Mach did come to have a genuine influence on the rise and progress of the special theory. Mach anticipated Bridgman by several years in what is now called the operational theory of the metrical concepts of physics. Einstein's special relativity, seen philosophically, is the classical application of

[1] A. Einstein, *Physikalische Zeitschrift*, **17**, 1916, 101–4.
[2] H. Dingle, *The Special Theory of Relativity*, p. 1.

Bridgman's theory. The matter is sufficiently well summed up in these two quotations:

> Let us examine what Einstein did in his special theory. In the first place, he recognised that the meaning of a term is to be sought in the operations employed in making application of the term.[3]
>
> It is easy to see which lines of Mach's thought have been particularly helpful to Einstein. The definition of simultaneity in the special theory of relativity is based on Mach's requirement, that every statement in physics has to state relations between observable quantities.[4]

Mach, Einstein and Bridgman share this insight that there are no operations for a metric of absolute displacement, absolute time and absolute velocity. So these concepts are irrelevant to physics and we must do without them.

Michelson and Morley, and their successors, made a very curious observation concerning the relative velocity of light, that is, the only kind of velocity of light which can conceivably be measured. They found that it takes light just as long[5] to pass between two rigidly joined parallel mirrors, however the mirrors may be set with respect to the motion of the earth. In other words, the velocity of light relative to a rigid framework, moving in any way through what used to be called the ether, always turns out to have the same constant value however the framework is moving or is orientated. The relative velocity of light is constant. The special theory of relativity is best approached from this observation. The observation itself has been found difficult to repeat and confirm, but it has been accepted by the best recent opinion.[6] Bondi has recently pointed out that the Michelson–Morley experiment is in fact no longer the true empirical basis of the special theory of relativity. He notes, for example, that the increase of mass with velocity and the equivalence of mass and energy are indicated by the special

[3] P. W. Bridgman, *Albert Einstein*, ed. P. A. Schilpp, I, New York: 1959, p. 335. [4] P. G. Frank, ibid., p. 272.

[5] Strictly, they saw interference fringes, and they did not measure time (H. Dingle).

[6] R. S. Shankland, S. W. McCuskey, F. C. Leone, G. Kuerti, 'New Analysis of the Interferometer Observations of Dayton C. Miller', *Reviews of Modern Physics*, **27**, no. 2, 1955, pp. 167–78. I am indebted to Dr G. J. Whitrow for directing my attention to this review.

theory, and that these effects have been obtained and studied with great experimental precision.[7]

In the *Optics*, although Mach gives an account of the Michelson refractometer, he omits altogether any mention of the famous negative experiment.[8] Perhaps this would have been rectified in the projected second volume, had Mach lived to write it. In the preface to the *Optics* as we have it, Mach disclaims to be "a forerunner of the relativists".[9]

7. Inertial frames

In a discussion of the collision between two perfectly elastic balls (1703) Huygens made use of the conception of the inertial frame.[1]

Huygens begins with the fact that if two equal perfectly elastic balls moving with equal and opposite speeds v (i.e., more precisely, with velocities v and $-v$ respectively) collide, they then separate again with equal and opposite speeds v (or with velocities $-v$ and v respectively). This is as Mach points out an experimental fact or law; it is not, as Huygens called it, an assumption, although the usual processes of refinement have been applied to the experiential basis of the law. Huygens then imagines that the two balls move at velocities v and $-v$ *relative to a boat* in which they move, and that the boat itself moves with velocity v. *From the point of view of a man in the boat*, the balls are supposed to move in exactly the same way as before; they separate therefore with velocities $-v$ and v *relative to the boat*. In this supposition Huygens makes a further appeal to experience; he exhibits insight into what are now called inertial systems or, if preferred, he makes a bold application of Newton's first law. The velocities $-v$ and v relative to the boat are 0 and $2v$ *relative to the fixed stars* or to the river bank. Relative to the fixed stars, the velocities before the collision were $2v$ and 0; again therefore, the two balls exchange their velocities as a

[7] H. Bondi, *Assumption and Myth in Physical Theory*, Cambridge: 1967, pp. 24–9.

[8] The old-fashioned way of describing it: It is impossible to detect the velocity of the earth through the ether.

[9] O., p. viii.

[1] M., pp. 403–15.

result of the collision. The diagram makes Huygens' result quite clear.

Velocity of boat $= +v$ (i.e. →).

FIG. 29

Velocities (rel. to boat):	$+v$	$-v$;	$-v$	$+v$.
Velocities (rel. to stars):	$+2v$	0;	0	$+2v$.

Finally the boat is, as it were, annihilated; it is assumed that the velocities $2v$, 0 are exchanged for 0, $2v$ on *terra firma*. Huygens easily generalises the result, and proves, by an extension of the same method in which the velocity of the boat is independent of the velocities of the balls, that on impact the balls exchange their velocities, whatever the values of these may be.

The significant step in Huygens' account is the annihilation of the boat; what is 'all right' for the boat is 'all right' for the bank. But the interesting question, perhaps not answerable, then arises: how 'all right' is the original empirical rule for the bank and any other inferences about the universe relative to the bank? So Albert Einstein and his colleague Leopold Infeld write with an exemplary caution which one dare hardly ignore:

. . . the so-called *Galilean Relativity Principle: if the laws of mechanics are valid in one c.s.*, [*coordinate system*] *then they are valid in any other c.s. moving uniformly relative to the first.*

If we have two c.s. moving non-uniformly, relative to each other, then the laws of mechanics cannot be valid in both. "Good" coordinate systems, that is, those for which the laws of mechanics are valid, we call *inertial systems*. The question as to whether an inertial system exists at all is still unsettled. But if there is one such system, then there is an infinite number of them.[2]

Equally justified are Bondi's candid comments:

Since the days of Galileo and Newton we have regarded velocity as relative but acceleration as absolute. This is difficult to understand, but there doesn't seem to be any simple way out. . . . There is nothing more difficult to grasp about relativity than Newtonian relativity—that there are inertial observers, that one of them is as

[2] A. Einstein and L. Infeld, *The Evolution of Physics*, Cambridge: 1938, pp. 165, 166.

good as any of the others, but that acceleration is something quite different.[3]

The ordinary student of classical physics, as he carries out his experiments in the laboratory, simply assumes, as Galileo, Newton and Huygens did, that displacements and velocities relative to the walls of the laboratory, or to the river bank, satisfy the Newtonian principles, or that this is at least nearly true. Indeed, if it were not nearly true, how could the principles, which are in part empirical, have been discovered? Newton saw that the concepts of absolute space, absolute time and absolute motion can jointly determine an ideal inertial frame. To avoid this metaphysical speculation Mach suggests that we had better take the stellar framework as the best actual inertial frame available. So Bridgman sees Mach's contribution to relativity in these terms:

> Not only does nature provide us with a unique operation for measuring interval, but it also provides us with a unique frame of reference, namely a frame fixed with respect to the stellar universe, as was pointed out by Mach.[4]

So, it seems, we no longer refer to a frame which is a mathematical construction or a philosophical idea; the stellar frame is a thing made out of matter, a physical or natural object. According to Bridgman, Mach was entirely correct in thinking that the law of inertia necessarily requires a reference to this material reference frame. Mach as it were *discovered* Descartes' axes for him—the only axes which have ultimate *physical* significance. This is a correct reading of Mach as far as it goes. The spin of the water in Newton's bucket relative to the stellar frame produces what Bondi calls an absolute acceleration in it. This acceleration is computed by means of the well-known quotient of the square of the speed by the radius; multiplied by a mass coefficient an absolute force is derived. At any rate, the force is absolute enough to produce the concavity of the water surface. This interpretation of the facts avoids the postulate of an absolute spin of the water in absolute space through absolute time, which seems to be meaningless.

[3] H. Bondi, *Assumption and Myth in Physical Theory*, pp. 20, 21.

[4] P. W. Bridgman, *Albert Einstein*, ed. P. A. Schilpp, I, p. 351.

But when Mach goes on to consider that we do not in fact know whether it is the bucket which is spinning, and that it could be the stellar frame, then Bridgman's comment about the fixed frame appears superficial. Mach's *complete* treatment of the bucket problem and his *overall* critique of Newton's dynamics amounts to an invitation to dispense with the idea of the inertial frame altogether. A long study of Mach's writings suggests to me that Mach is a true herald of general relativity, whether he cared to be seen in this light or not. His relationship to the special theory is less significant.

What is meant by dispensing with the inertial frame is made clear by one of Einstein's favourite fictions—the man in the box. I follow here Sciama's very clear exposition.[5] One begins with a Newtonian account of the supposed events. The diagram is Sciama's, but I have added a few stars to provide a stellar frame. In Fig. 30a, the man in the box is at rest in remote

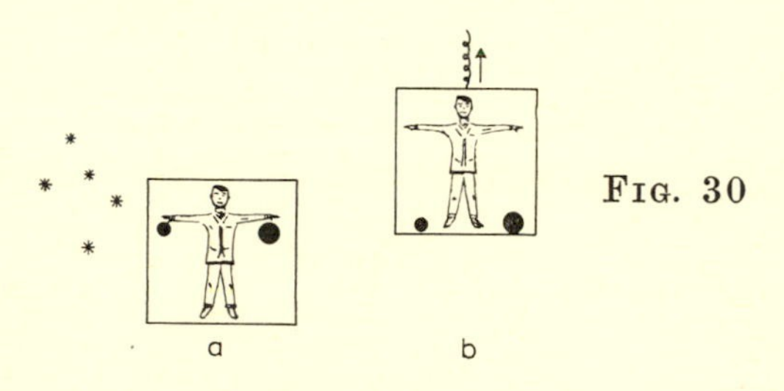

Fig. 30

absolute space, or in relation to the stars. There are two balls of unequal mass just below the man's outstretched hands. In accordance with Galileo's principle the man feels weightless, he is stationary with respect to the stars and remains so, and the balls do not 'fall'. Then a cosmic demon puts a chain on the box and pulls it in the direction shown relative to the stars. The force exerted by the demon is given by the product

$$(\text{mass of man} + \text{mass of box})\ g$$

where g is the acceleration due to terrestrial gravity, and is here relative to the stars as usual (Fig. 30b). The man suddenly acquires weight and he feels his weight pressing on the floor of

[5] D. W. Sciama, *The Unity of the Universe*, London: 1959, pp. 104, 105.

the box. The two balls stay where they are relative to the stars, but in the box they look as if they are falling *together*, with the acceleration g to the floor of the box. Now Einstein takes it as self-evident that the man can take his box as his 'frame'. He cannot know that it is not an inertial frame. He cannot know that an inertial force is being exerted on his box, and that no force is being exerted on the balls. He is perfectly free to regard the balls as falling relatively to his box with the common acceleration g, as balls fall on the earth. From his point of view, the forces on the balls could be gravitational. This is a most interesting exploitation of the simple fact known to Stevinus and Galileo, that different masses fall together. Both inertial and gravitational forces are proportional to mass. Einstein takes the view that " . . . *there is no criterion whatsoever by means of which an inertial force can be distinguished from a gravitational one*".[6]

That this is a fair account of Einstein's own starting-point in general relativity is shown by his own words:

Now it came to me: The fact of the equality of inert and heavy mass, i.e., the fact of the independence of the gravitational acceleration of the nature of the falling substance, may be expressed as follows: In a gravitational field (of small spatial extension) things behave as they do in a space free of gravitation, if one introduces in it, in place of an "inertial system", a reference system which is accelerated relative to an inertial system.

If then one conceives of the behaviour of a body, in reference to the latter reference system, as caused by a "real" (not merely apparent) gravitational field, it is possible to regard this reference system as an "inertial system" with as much justification as the original reference system.

So, if one regards as possible, gravitational fields of arbitrary extension which are not initially restricted by spatial limitations, the concept of the "inertial system" becomes completely empty. The concept, "acceleration relative to space," then loses every meaning and with it the principle of inertia together with the entire paradox of Mach.[7]

Einstein's interpretation of the man in the box is not without

[6] Ibid. [7] A. Einstein, *Albert Einstein*, ed. P. A. Schilpp, I, pp. 65–7.

its own peculiar difficulties. For example, the Newton–Berkeley–Eddington principle can be applied again, and its application seems to blunt the force of Einstein's reasoning. If, in Fig. 30b, the man knows that he has had to engage the services of the cosmic demon, he is not likely to think that his suddenly acquired weight and his sense-perception of the falling balls have come about because of his return to earth. He can see for himself that the earth is not there through the glass floor of the box and he interprets the force ↑ as inertial. If alternatively, he has not called in the demon, but has the same experiences together with a glimpse of the earth quite near, he again gives a purely Newtonian interpretation. For indeed action-at-a-distance, in Newton's dynamics and gravitation physics, is noticeable-action-when-the-distance-is-not-too-great.

With further reference to the man in the box, Einstein asks us to consider that the two unequal balls, 'falling' accelerated together, may be interpreted by an acceleration of the box relative to the stars; this is an alternative to the classical view that the balls are subjected to gravitational forces proportional to their masses. This amounts to eliminating the notion of gravitational force. As the balls fall there is no way of computing the forces to which we were wont to think they are subjected. Bondi's way of putting this is very helpful. We used to think of acceleration as absolute, of velocity as relative. According to general relativity, we no longer need to draw this distinction. But if so, force as the product of mass and acceleration, is no longer absolute. Bondi states simply that "Gravitation is not a force",[8] and Eddington commented, half a century ago, that "force is purely relative."[9]

It is not easy to reconcile faithfulness to Mach's first principles with this consequence of general relativity. To Mach there is nothing more actual than a force; force is an 'O' concept and it enjoys the simplest direct relationship to sense-perception. Mach must have felt that if this position is in any way threatened by general relativity, so much the worse for general relativity. The problem is elucidated, but not I think solved, by

[8] H. Bondi, *Assumption and Myth in Physical Theory*, p. 58.

[9] A. S. Eddington, *Space, Time and Gravitation*, Cambridge: 1920, p. 76.

this comment of the astute critic, N. R. Campbell: "... statical force is an experimental concept and a magnitude, dynamical force is a hypothetical idea."[10] We can roughly 'sense-perceive' statical force and also acceleration; these are both 'O' concepts. We cannot even roughly 'sense-perceive' dynamical force as the mass × acceleration product. This, I take it, is Campbell's meaning. Yet when I whirl a massive ball around in a plane circle on a horizontal floor, I can actually feel the elastic metal wire pulling my hand outwards; I can calculate the force on the ball, both dynamically from Newton's second law and the formula for central acceleration, and also from the extension of the wire and using the results of a Hooke's law statical experiment. In fact the notions of dynamical force, statical force and sense-perception are all satisfactorily welded into an excellent instance of the doctrine of 'return' discussed in an earlier chapter. It seems to me, therefore, that if there is some logical or philosophical difficulty in reconciling general relativity with the sense-perception of force and *any form* of the metrical concept of force in Newton's mechanics, the same difficulty will arise for *any other form* of the Newtonian concept of force. This is a pedantic way of expressing the satisfactory unity and integration of Newton's system. A part cannot be challenged without the whole being challenged.

I conclude with two quotations from Burniston Brown's paper:

> Einstein says: "the purpose of mechanics is to describe how bodies change their position in space with time." But this, of course, is the purpose of kinematics. Mechanics is the science dealing with forces, but forces cannot be put into geometry. . . . The simple fact remains that the gravitational attraction between bodies at rest is a centripetal force—a force tending to a centre—as Newton said long ago, and the force of inertia is not. This is a fundamental physical difference which cannot be brushed aside by fumbling about in closed chests or manipulating the root of minus one.[11]

[10] N. R. Campbell, *Foundations of Science*, New York: 1957, p. 560. First edition dated 1919.

[11] G. Burniston Brown, 'Gravitational and Inertial Mass', *American Journal of Physics*, **28**, no. 5, May 1960, pp. 482, 483.

8. The Mach Principle (Einstein)[1]

Mach's comment on the spinning bucket with walls a mile thick, and his reference to the universe spinning round the bucket, lead to the Mach Principle (Mach): the possibility that the matter of the whole cosmos could affect terrestrial dynamical systems is left open. How could the remote matter of the distant nebulae affect our experiments here? Presumably by forces of the gravitational kind, for gravitation is a central force between material bodies. The forces on the water in Newton's bucket are inertial; they are calculated from the acceleration which characterises circular motion and Newton's second 'law'. Mach's open possibility leads to a possible unification of mechanics; more precisely, to the removal of the distinction between gravitational and inertial forces.

Such presumably is the bridge between the Mach Principle (Mach) and the Mach Principle (Einstein). The reader is reminded that Mach himself neither built nor crossed this bridge. Three statements of the Mach Principle (Einstein) are given. They are not identical:

> . . . Mach's principle, was perhaps put most beautifully by Einstein himself when he said that in a consequential theory of relativity there can be no inertia of matter against space, only an inertia of matter against matter.[2]
>
> . . . Mach's principle, according to which the inertia of a body is due to the presence of all the other matter in the universe.[3]
>
> A century and a half later [i.e. than Berkeley's *De Motu*] Mach put forward similar ideas, suggesting that inertial forces are produced by motion with respect to distant matter.[4]

9. General relativity and other theories of gravitation

The general relativity theory of Einstein is a theory of gravitation. Einstein replaces 3-dimensional Euclidean space by a non-Euclidean space whose local curvature is determined by the

[1] More justly, but less elegantly, this should be: Mach Principle (Expositors of Einstein). Note suggested by Professor Dingle.

[2] H. Bondi, *Assumption and Myth in Physical Theory*, p. 70.

[3] E. A. Milne, *Modern Cosmology and the Christian Idea of God*, Oxford: 1952, p. 69.

[4] G. Burniston Brown, loc. cit., p. 481.

density of matter. Consequently he can dispense with the Newtonian conception of masses exerting forces on each other 'at a distance', and indeed the tendency of the theory is to eliminate forces from physics altogether, as we have seen. Mach teaches us that *the space of our common experience has three dimensions* and that *there are forces in nature*. It is easy to understand then why Mach dislikes relativity.

Einstein's general theory has been called 'a field theory'. Attempts have been made to develop non-field theories of gravitation. For example, Burniston Brown has suggested that there may be a mechanical force between moving masses, somewhat like the mechanical force between linear conductors carrying electric currents. "Let us suppose", he writes, "that gravitational force varies with relative motion in the same way as macroscopic electrical force."[1] The authors of a new theory of gravitation announced in 1964[2] claim that their theory is a 'particle' or 'action-at-a-distance' theory, and not a 'field' theory. This claim has not however been accepted universally.[3]

[1] G. Burniston Brown, loc. cit., p. 481.

[2] F. Hoyle and J. V. Narlikar, 'A New Theory of Gravitation', *Proc. Roy. Soc.* A, **282**, 1964, pp. 191–207.

[3] S. Deser and F. A. E. Pirani, 'Critique of a New Theory of Gravitation', *Proc. Roy. Soc.* A, **288**, 1965, pp. 133–45.

7 The intellectual element in science

1. Introduction

The intellectual element in science is not something added on *after* an experiment. Newton's *hypotheses non fingo* expresses an important truth, but it would be an absurd error if it were taken to mean that any part of the process of science can take place without the action of the human mind. *Before* Galileo can begin his investigation on falling bodies, his mind is informed by preconceived opinions or prejudices. Were this not so he would have neither motive to experiment at all nor chance of success:

> Without some preconceived opinion the experiment is impossible, because its form is determined by the opinion. For how and on what could we experiment if we did not previously have some suspicion of what we were about? . . . The experiment confirms, modifies, or overthrows our suspicion.[1]

This affirmation by Mach is the more striking because the technical term 'hypothesis' and the philosophical jargon 'hypothetico-deductive method' are avoided.

Mach considers it important to trace the metrical concepts of physical science back to sense-perception. The line back may be short (from force, temperature) or long (from energy, entropy) but it is always a line of thought. At the same time, he is aware that the sense-perception by itself requires intellectual interpretation. The mere succession of images on the retina does not enable Galileo to *perceive* a falling stone. The psychological equation for the simplest experience is this:

$$\text{sensation} + \text{mind} = \text{sense-perception.}^2$$

[1] M., p. 161. [2] P., p. 35.

From the pushes and pulls felt by primitive man long before science achieved self-consciousness, right on to the general theory of relativity and quantum mechanics, the human mind is at work. There is no science at all without thought. The intellectual element in science is like the yeast in bread.

This being so, it is hardly possible to give a good account of the intellectual element apart from the rest of science. This accounts for the fact that Mach's best interpretations of science occur in his three textbooks of physics.

The passage which follows may be taken as a survey of the subject as a whole:

As a natural inquirer, I am accustomed to begin with some special and definite inquiry, and allow the same to act upon me in all its phases, and to ascend from the special aspects to more general points of view. . . . I was obliged to proceed in this manner for the reason that a theory of theory was too difficult a task for me, . . . I accordingly directed my attention to individual phenomena: the adaptation of ideas to facts, the adaptation of ideas to one another, mental economy, comparison, intellectual experiment, the constancy and continuity of thought, etc. In this inquiry, I found it helpful and restraining to look upon every-day thinking and science in general, as a biological and organic phenomenon, in which *logical thinking assumed the position of an ideal limiting case.*[3]

2. The metrical concept and alternative concepts

In the passage quoted above Mach distinguishes between the

adaptation of thought to fact

and the

adaptation of thought to thought.

This is a useful initial analysis of the theoretical element of science. The metrical concept is the most important example of theory of the first kind: the adaptation of thought to fact. Whether the 'line' back to factual experience is long or short, complex or simple, it must be traceable.[1] The factual experience

[3] M., pp. 592–3.
[1] A., p. 364.

is capable of analysis into sense-perceptions. One wonders if Mach might have distrusted some of the concepts of the physics of 1971 on the grounds that the 'lines back to experience' are too long or too ill-defined. What length of 'line', what lack of definition of 'line', would Mach have tolerated?

The metaphor of the 'line' is an alternative to the metaphor of 'return', given by Robert Hooke and briefly illustrated in the first chapter of this book. In classical dynamics, force ('O') is defined in terms of mass ('non-O'); in classical thermodynamics, temperature ('O') is defined in terms of energy ('non-O'). The two definitions are important *parts* of the return journey (←)

$$\text{sense-perception} \rightleftarrows \text{'non-O' concept.}$$

Mach's theory of the metrical concepts is thus written into standard classical theoretical physics.

According to Mach, the kinds of concept employed by the physicist, and the kind of theoretical system built up from the concepts, are "in some degree conventional and accidental".[2] Three examples of this are here given. The first one is not given by Mach, and has been devised to reduce the idea to its simplest terms.

On the original Celsius scale of temperature, water boils at 0°E and the temperature of melting ice is 100°E, the letter E being used to distinguish the first Celsius scale from the more familiar C scale. A person familiar with the E scale might well have been led to think of cold water as containing 'more cold'[3] than warm water, particularly as its temperature is higher. Without the excuse of the E scale being familiar to them, simple people have been known to shut the doors in order to keep the cold out. Now let us solve a simple problem, using the Black conceptual system, and also a logical alternative. The two solutions are set out step by step in parallel columns. *Problem:* What is the temperature of the mixture, when 30 g of water at 80°C is mixed with 60 g of water at 40°C?

[2] M., p. 316 (my translation).

[3] H. Dingle, *The Scientific Adventure*, London: 1952, pp. 21–9.

I	II
30 g water at 80°C	30 g water at 20°E
is mixed with	is mixed with
60 g water at 40°C.	60 g water at 60°E.
Let mixture be at t°C.	Let mixture be at t°E.
Heat lost by warmer water	Cold gained by warmer water
$= 30\ (80 - t)$ *calories*.	$= 30\ (t - 20)$ *frigories*.
Heat gained by cooler water	Cold lost by cooler water
$= 60\ (t - 40)$ calories.	$= 60\ (60 - t)$ frigories.
By hypothesis,	By hypothesis,
Heat lost $\equiv$ heat gained.	Cold gained $\equiv$ cold lost.
$30\ (80 - t) = 60\ (t - 40)$,	$30\ (t - 20) = 60\ (60 - t)$,
whence	whence
$t = 53\frac{1}{3}$°C.	$t = 46\frac{2}{3}$°E.

Just before writing this, I was engaged in making tea. I was careful to warm the pot in the usual way. As I shook out the water, I felt I was getting rid of as much 'cold' as possible. One could quite easily learn to live in the alternative conceptual world.

The second example is taken from the early monograph, *The Conservation of Energy*.[4] In 1838, Riess designed an instrument which would transform electrical energy entirely into heat. The instrument consisted essentially of a fine metal wire passing through the air bulb of a gas thermometer. The wire was earthed. When an electrically charged ball, or a capacitor, was earthed through Riess's instrument, all the energy of the charged body was effectively degraded into heat. Let us first consider a set of operations, and interpret them in the usual way in the light of the traditional concepts of classical physics. Suppose a capacitor of capacity C carries a charge of Q. Then its energy, i.e. the mechanical work done in charging it or in bringing up the charge from infinity, is equal to $\frac{1}{2} . Q^2/C$. Now suppose the capacitor to be connected, *through the instrument of Riess*, to another capacitor of the same capacity C. The

[4] G., pp. 45–6.

charge Q will be divided equally, each capacitor now carrying $Q/2$. The energy of each jar will now be

$$\frac{1}{2}\cdot\frac{\left(\frac{Q}{2}\right)^2}{C}=\frac{1}{8}\cdot\frac{Q^2}{C}.$$

The total energy of the two jars together will therefore be

$$\frac{1}{4}\cdot\frac{Q^2}{C},$$

and half the original energy has been degraded into heat in the Riess instrument. Electricity has been conserved because

$$\frac{Q}{2}+\frac{Q}{2}=Q;$$

energy has been conserved because

$$\frac{1}{8}\cdot\frac{Q^2}{C}+\frac{1}{8}\cdot\frac{Q^2}{C}\text{ (electrical potential energy)}$$

$$+\frac{1}{4}\cdot\frac{Q^2}{C}\text{ (heat in Riess's instrument)}=\frac{1}{2}\cdot\frac{Q^2}{C}.$$

What is the point of going through this trite process? It is that *the actual character of the reasoning, and the main result of it, are at least in part determined by purely historical and accidental factors*. Mach invites us to consider how differently we might have regarded this problem, if Riess's instrument had been invented before Coulomb's torsion balance. Now this is not difficult to imagine. The Leyden jar was very early in physics (1746), and the measurement of heat quantities (1762) was pre-Coulomb (1785). It was, Mach considers, the use of Coulomb's torsion balance, which led to the familiar conviction that electricity is a conserved substance. Without that remarkable instrument, we should not have been so committed to the idea.

In any case, for the sake of the purely *methodological* question under discussion, it is quite fair to imagine the earlier invention of Riess's instrument. *We can therefore decide to measure electricity in terms of the heat generated when the body is completely discharged through Riess's instrument.* Now the experiment with the capacitor is reconsidered. Suppose the original energy (electrical potential energy) of the capacitor is E. The electrical charge on it, according to the new convention, is also E (in appropriate units). If the capacitor is completely discharged through the Riess instrument, the heat generated is also E. Now we connect the capacitor, through the Riess instrument, to an equal uncharged capacitor, and separate them afterwards. The heat generated in the Riess instrument is $E/2$. Now we discharge each of the two capacitors separately and completely through a Riess instrument. By doing this we find that each of the capacitors had the charge $E/4$. Of course the two further amounts of heat generated are each $E/4$ also. Let us now confine our attention to that part of the operation where the original capacitor is connected to an equal capacitor through the Riess instrument. There is no overall loss of energy for

$$E \text{ (electrical potential)}$$

$$= \frac{E}{4} + \frac{E}{4} \text{ (electrical potential)} + \frac{E}{2} \text{ (heat)}.$$

There is however a loss of electrical charge:

$$E \neq \frac{E}{4} + \frac{E}{4}.$$

In fact, in our new physics, electricity is destroyed when it produces heat, just as in the thermodynamics of Clausius, heat is destroyed when it does mechanical work. Mach comments:

The reason, therefore, why we have other notions of electricity than we have of heat, is purely historical, accidental, and conventional.[5]

Perhaps more interesting than either of these examples is the one drawn in the main from the *true* history of physics.[6]

A brief outline only is given here.[7] When Galileo began his

[5] G., p. 46. [6] M., pp. 305–17.

[7] J. Bradley, *Ernst Mach's Philosophy of Science*, pp. 471–9.

investigation on the law of falling bodies, he entertained two hypotheses, *viz.*

$$v \propto t$$

and

$$v \propto s.$$

For an incorrect reason, he judged the second hypothesis to be impossible. He was fortunate that the first hypothesis is the fact, and he proved it to be such by experiment. From

$$\frac{\mathrm{d}s}{\mathrm{d}t} = gt,$$

$$\frac{\mathrm{d}^2s}{\mathrm{d}t^2} = g$$

is easily derived. Progress towards the concept of force is suggested. For if

$$\frac{\mathrm{d}^2s}{\mathrm{d}t^2} = g,$$

$$m\frac{\mathrm{d}^2s}{\mathrm{d}t^2} = mg.$$

From this, Newton can easily construct the dynamics of a particle, in which major concepts are acceleration, force and momentum. The first law of motion for an isolated particle is seen as

$$\frac{\mathrm{d}(mv)}{\mathrm{d}t} = 0$$

or

$$mv = \text{constant}.$$

Now suppose Kepler had undertaken the initial research instead of Galileo. He would have known that

$$v \propto s$$

is entirely possible and intelligible, but he would have found that it is not the case. Not daunted he would quickly have reached the true law,

$$v \propto \sqrt{s}.$$

From this,

$$v^2 \propto s$$

and

$$mv^2 \propto ms.$$

There is an easy route this way to the dynamics of a particle, in which major concepts are work and kinetic energy. The first law of motion would be seen differently as

$$\tfrac{1}{2}mv^2 = \text{constant}.$$

A marble moving on a perfectly smooth horizontal table cannot gain or lose any of its kinetic energy as potential energy, because the table is horizontal; it cannot lose it as heat, because the table is smooth. The interest of this is that such alternative modes of interpretation did actually exercise the minds of the great physicists of the time. The whole of Mach's discussion[8] can be set out schematically, like this:

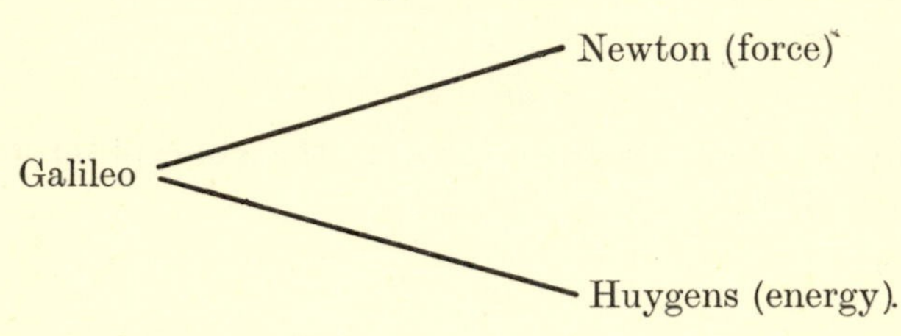

Such examples may suffice to show that physicists should take a tentative attitude towards their modes of interpretation and the theoretical concepts they have seen fit to employ.

3. Law and induction

In the formal discussion of deduction and induction[1] Mach follows Hume very closely. Like Hume he reduces causality and 'necessary connection' to mental familiarity or habit of mind. In one important respect, however, Mach goes beyond Hume. He realises that induction is an imaginative act of the mind, and he quotes this important passage from Whewell:

> The doctrine which is the *hypothesis* of the deductive reasoning, is the *inference* of the inductive process. . . . But still there is a great difference in the character of their movements. Deduction descends steadily and methodically, step by step: Induction mounts by a leap which is out of the reach of method. She bounds to the top of the stair at once; and then it is the business of Deduction, by trying each step in order, to establish the solidity of her companion's footing.[2]

[8] M., pp. 311, 312.

[1] E., pp. 304–19.

[2] W. Whewell, *The Philosophy of the Inductive Sciences*, II, London: 1847, p. 92.

If in fact what Whewell calls 'induction' is equated with 'the hypothesis' of the so-called 'hypothetico-deductive method', Whewell's words could be taken as a picturesque account of the main parts of that method. Induction is a mental act which lacks safeguards, it is insecure and fraught with risk. It is not properly to be called thinking.

Neither the syllogism nor induction are to be found at the source or spring of physical knowledge. Whatever they are, and whatever is achieved by them, they are not concerned in the creation of new physical knowledge. This is the plain meaning of Mach's words:

> Consequently neither the syllogism nor induction can create new knowledge. But they do guarantee the establishment of freedom from contradiction between different parts of knowledge, they set out clearly the connection between these parts, they direct our attention to the various sides of an individual insight, and they teach us to recognise a given insight as one and the same, even when it is presented in different forms. It is therefore clear that the investigator's proper source of knowledge must lie elsewhere.[3]

And the proper source must be the sense-perceptions or 'elements' of experience.

Mach expounds the rather tedious and old-fashioned doctrine of perfect and imperfect induction.[4] If *all* the B's have been examined, and if every B is found to be an A, then the induction

All B's are A,

is a *perfect* one. If, however, not all the B's have been examined, or if it is not sure whether all the B's have been located, and yet nevertheless we affirm that

All B's are A,

as before, then we have made an *imperfect* induction. It is the imperfect induction which in general is relevant to the quest called science, but, as Mach promptly comments, it "has *no logical justification whatever*."[5] This blunt statement puts Mach on the side of modern writers like K. R. Popper and R. Carnap,

[3] E., p. 312. [4] E., p. 308. [5] E., p. 308.

and is in plain opposition to John Stuart Mill who made out that imperfect induction is a logical process.

Mach follows this by a brief statement of Hume's doctrine. We may *want* B to be A, we may *expect* a new B to turn out to be an A, and undoubtedly we are led to such expectation by the powerful influence of *psychological association or habit*. Although we ought not to write

B is A,

we feel we might venture as far as

B may be A.[6]

When the metal rhenium was discovered in 1925, it was soon identified as a 'transitional metal'. Chemists were immediately prepared to discover that rhenium would form complex anions. The preparation of the perrhenates confirmed the expectation. There is no doubt that this process of imperfect induction is extensively used, and is of the greatest value in both scientific investigation and everyday life. Mach, following Hume, considers that, to assert that only a *psychological* account of it can be given, need not be a form of disparagement. It is countless imperfect inductions which furnish the mind of the investigator with the mysterious quality called 'insight'. Although in a formal sense imperfect induction holds within it no extension of knowledge, nevertheless it represents an anticipation of new knowledge:

> It is true, on the other hand, that imperfect induction *anticipates* an extension of knowledge. At the same time it has, included within itself, the risk of error. So, it is *a priori* quite definite, that in the first place the imperfect induction is to be put to the test, and corrected or entirely rejected.[7]

This expectation or anticipation is strengthened through frequent repetition. In the classical illustration of Hume, every time a ball is seen to convey motion to another ball, the mind of the observer is being trained or educated to expect similar events in the future. Mach states this in a powerful and interesting way,[8] but he makes no essential advance beyond the

[6] "B sei A". [7] E., p. 309. [8] W., p. 383.

position of Hume. His comment on "extreme values" is worth quoting:

The variations of a phenomenon with the attendant circumstances stimulates the desire to find out how the former is affected by *extreme values* of the latter.[9]

This is exemplified by Kepler's brilliant work on refraction. Kepler made a complete survey of the phenomena of the refraction of light from glass to air and his attention was directed to a gap in his knowledge (Fig. 31). What can possibly happen to

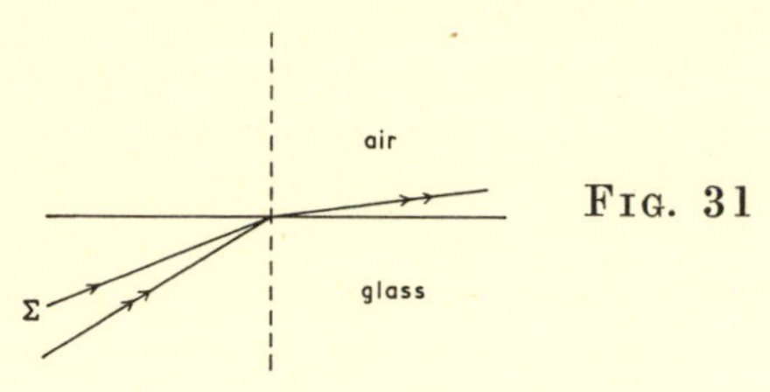

FIG. 31

the ray Σ incident at an angle greater than the critical angle, and which cannot therefore proceed as a refracted ray if the sine law is to be obeyed?[10] In an important footnote, Mach pillories the dreadful logic of many of the textbooks:

$\sin r$ (where r = angle of refraction) > 1

which is impossible;

refraction cannot therefore take place;

therefore the ray must be reflected.

The ray Σ *could* have changed its degree of refrangibility and been transmitted as X-rays, its energy *could* have been degraded into heat energy and so on. Only experience can tell us what does happen; nor, wise after the event, may we pretend that we reached the knowledge by a logical process.

Mach returns frequently to the example of the law of refraction, that

$$\frac{\sin i}{\sin r} = \mu$$

where i = angle of incidence,

r = angle of refraction

and μ = a constant for the two media.

[9] E., p. 217. [10] O., pp. 31, 32.

When a graph of sin i and sin r is made, a set of points falling on a straight line is revealed. Here, in the most undeniable form, is the problem of induction: the line is a mathematical continuum setting forth a precise relationship between values of sin i and sin r, and there is no finite number of values. Yet the law is erected on the basis of a finite number of operations. The law is nothing more, says Mach, than a "comprehensive and condensed report about facts."[11] In a sense, the law contains less than the facts; the facts are about pins and geometrically constructed lines whereas the law is about rays of light and a trigonometrical function. As Mach puts it, the law "does not reproduce the fact as a whole" and much is "intentionally or from necessity omitted".[12] But it is equally true that, in a different sense, the law contains much more than the experienced facts, for the facts are generalised to a number larger than any finite number however large; the points become a line. Strictly we cannot experience laws and there are, as Mach says, no laws to be found in nature: "In nature there is no *law* of refraction, only different cases of refraction."[13] It is fair to ask, then, 'Why do we have laws?' According to Mach, we have the law of refraction so that we can produce in thought any specific case of refraction, by simply recollecting the value of μ and performing a simple calculation. We could not possibly remember an infinite array of all possible i and r values. The law is, above all else, a means to economise thought.

Again, to save the labour of instruction and of acquisition, concise, abridged description is sought. This is really all that natural laws are.[14]

Few (Poincaré is an exception) would entirely agree with Mach; the truth about laws is *more* than this, but I believe it *contains* this.

I have said that the truth about laws is more than Mach gives us. The obvious gap in Mach's account is the puzzling question of the future reference of laws. As early as 1892 Karl Pearson was aware of this problem:

That a certain sequence has occurred and recurred in the past is a

[11] P., p. 193. [12] Ibid. [13] M., p. 582. [14] P., p. 193.

matter of experience to which we give expression in the concept *causation*; that it will continue to recur in the future is a matter of belief to which we give expression in the concept *probability*. . . . Science for the past is a description, for the future a belief; it is not, and has never been, an explanation, if by this word is meant that science shows the *necessity* of any sequence of perceptions. . . . No sequence of sense-perceptions involves in itself a logical necessity . . . it has become for us a routine by repeated experience. . . . Hence, when we are speaking of the sphere of perceptions we must remember that provable is ultimately the same word as probable. The association of the two words does not therefore seem without profit; and the etymology may after all serve to remind us of the character of our knowledge in the field of perception.[15]

As Harold Jeffreys comments:

Mach missed the point that to describe an observation that has not been made yet is not the same thing as to describe one that has been made; consequently he missed the whole problem of induction. This was taken up by Karl Pearson in *The Grammar of Science*, . . .[16]

This is an overstatement. But Mach certainly 'misses' an important part of the problem of induction, namely that part which refers to the future reference of the law or general statement.

R. B. Braithwaite has adversely criticised Mach's view for the same kind of reason. If the law is merely to help us to describe nature economically, then we must also be able to describe the future and the unobserved.[17] To assert that one can describe the future is nonsense.

4. Hume and Keynes

Hume's account of the billiard balls and the eggs which are "so like" reduces the status of a part of the activity called science from the logical to the psychological. The strength of Hume's argument can now be more fairly judged. If an induction is both universal and possessing future reference, it is difficult to

[15] K. Pearson, *The Grammar of Science*, pp. 99, 112, 121.
[16] H. Jeffreys, *Scientific Inference*, Cambridge: 1957, pp. 15, 16.
[17] R. B. Braithwaite, *Scientific Explanation*, New York: 1960, p. 348.

see by what logical process it could possibly be derived from experience, which must be particular and confined to the present and past. Kant accepts this, but finds room for the logical reason in the application of the general categories of the mind. The Kantian categories are not empirical, and not hypotheses in any sense.

In an important book, John Maynard Keynes suggested another way in which reason could be restored to the practice of science. Any future phenomenon may or may not occur, but Keynes believes that the probability of its occurrence may be estimated, in principle at least, by a rational process. If a physicist predicts a future event as certain to occur, his mode of procedure must be irrational; if he is content to judge the probability of the occurrence his mode of procedure may be rational. In this sense, the notion of probability restores rationality to physical science.

> I have described Probability as comprising that part of logic which deals with arguments which are rational but not conclusive.[1]

In the same book Keynes, at least partially, accounts for the need of a large number of instances in an inductive process. Why must Hume try many eggs before he can reach the conclusion that 'eggs are nourishing'? It could be that very small eggs cause pimples or that brown eggs lack a certain vitamin. It is because the eggs are not quite alike, although they may be "so like", that the "long course of experience" is needed. After the experiments, one can say that although an egg may be this, that, or the other, yet 'eggs are nourishing'. The differences in the samples are called by Keynes the negative analogy.

Keynes also draws a useful distinction between the *known* and the *total* negative analogy.[2] When an observer turns to the second or third instance, he may be adding to the negative analogy without knowing that he is doing so. Evidently the negative analogy between instances is one of the factors which will determine the estimated probability of the occurrence of some future event.[3]

[1] J. M. Keynes, *A Treatise on Probability*, London: 1943 (first edition 1921), p. 217.

[2] Ibid., p. 223. [3] Ibid., p. 233.

5. Perfect objects and thought-experiments

The themes of law, induction, perfect object and thought-experiment are closely related. This has been indicated briefly already, in connection with Galileo's proof of the law of inertia (Figs. 15, 17). For contingent practical reasons the experiments with the double inclines and pendulums come to an end. They are then continued, as a thought-experiment, by the mind of the investigator, who assumes that nature exhibits a degree of continuity. The final result is a generalisation about something which has never been and can never be exhibited to sense-perception. The making of the generalisation includes an induction. At the same time, the generalisation or law is truly empirical. The paradox that, although the generalisation is empirical yet no example of it can be demonstrated, can only be resolved by noting that the objects and environment which are required for the law to obtain, are *idealised objects*, also known as *perfect objects*. At the same time, the idealised objects are not pure creations of the mind; if they were, laws about them would have no interest to the physicist. They are objects with the same kind of attributes as actual objects; the attributes however are 'heightened' or 'idealised'. As an example, the smooth horizontal plane on which the motion of a sliding object truly would demonstrate Galileo's principle, is in fact like the surface of a frozen pond.

It is convenient, but not entirely correct, to think of this typical kind of thought-experiment, as one in which the thought-experimenting occurs *after* the experiment. This is artificial for, of course, no experiment is undertaken without previous thought. Indeed, as Mach points out in the important passage quoted earlier[1] in this chapter, a man without some "preconceived opinion" would be in no position to undertake an experiment at all. Mach uses the term 'thought-experiment' to cover almost the entire intellectual element of science. This is a natural usage, although it can confuse the reader not well versed in Mach's work. Consistently, therefore, Mach also describes the thought which is prior to the experiment as 'thought-experiment':

[1] M., p. 161.

But the thought-experiment is also a necessary *inescapable condition* of the physical experiment. Any experimenter or investigator must have a scheme worked out in his mind, before he translates the same into action.[2]

And without such prior intellectual exercise there can be no experiment. Galileo's hypothesis that

$$v \propto t$$

for a freely falling body is thus a legitimate example of a thought-experiment.[3] It is difficult to deny that such a hypothesis is an experiment of the mind, leading to further experiments of the mind. Indeed, what is normally called Galileo's deduction, that

$$s \propto t^2,$$

could be described in this new way. Since then Mach draws no hard and fast distinction between a hypothesis and a thought-experiment; since, further, he considers that any experiment must be preceded by thought-experimenting; then it follows that Mach takes the view that any experiment is preceded by some kind of hypothesis. This was his opinion in 1905.

It is entirely artificial to think of the thought-experiment as either succeeding or preceding the experiment. Mach gives us a true picture of science as islands of experiment in a sea of thought. Yet, at least in one passage where he suggests that thought-experimenting "leads to theory",[4] he finds it convenient to distinguish between the thought-experiment as a comparatively simple intellectual act and theory as something more complex. If thought-experimenting leads to theory, the process is reversible, for *the somewhat incomplete theory leads back again to further thought-experiment*. The investigator is in the position of one who seeks to recollect something, one who is aware of a gap in his knowledge, a man faced with an incomplete jig-saw puzzle.[5] The complete jig-saw puzzle represents the theory, conceptual system, or classification. At the stage when the picture is not complete, which in practice means almost always,

[2] E., p. 187. [3] Algebraic symbols used as before (p. 101).
[4] E., p. 286. [5] P., pp. 275–8.

all experimenting, whether actual or intellectual, is directed towards completing the picture. In this sense Mach's statement, that the thought-experiment leads to the theory, is correct. At the same time, the incompleted theory suggests to the imagination of the investigator what is missing, what it would be most desirable to find. This is the reciprocal process; the theory constrains the worker to further experimenting, again of both kinds.

Mach's own example of this is Faraday's discovery of electromagnetic induction.[6] Oersted had found that a current in a wire gives rise to a magnetic field, the direction of the field being at right angles to the wire. To the powerful mind of Faraday, this carried the implication of an undiscovered phenomenon, the production of a current in a wire by means of a magnetic field. The desired whole was perhaps, in the first place at least, a two-piece jig-saw picture, and Oersted had revealed one of the pieces only. Faraday's actual discovery was that a constant magnetic field, however powerful, produces no current in a neighbouring conducting circuit; if however the magnetic field is changing then a current is induced. It is unlikely that the intellectual experiment, even of a Faraday, suggested the true fact in all its subtlety. Once the actual experiment is done, the two-piece picture becomes itself incomplete. It is not quite what was expected, and it is not tidily circumscribed.

Mach powerfully uses the correlative ideas of the thought-experiment and the perfect object to wed the two sciences of geometry and physics. The earlier discussion of Euclid's 'theorem four' (Fig. 8) is easily translated into different terms. One imagines oneself cutting out the triangle ABC either from the paper with real scissors, or from two-dimensional space with 'thought scissors'. Euclid's proof could be described as an *operational thought-experiment.* It is important to note that the mind does not merely *see,* in the eye of the mind, the two triangles and their attributes; it imagines its own body and hands carrying out a "set of operations".

The combined action of the sensuous imagination with idealised concepts derived from experience is apparent in every geometrical deduction.[7]

[6] E., pp. 218–19. [7] E., p. 384.

An examination of Euclid's proof[8] suggests that it should not strictly be described as logical. It is certainly 'synthetic', in Kant's sense. The theorem is valid for perfect objects or idealised triangles only; Euclid may well have thought of these as Platonic 'ideas'.

To Mach, Huygens' treatment of elastic collisions and Carnot's reversible engine cycles are *advanced thought-experiment fictions*. Mach's discussion of the thought-experiment is untidy and informal; he tends to equate the thought-experiment with any kind of theory at all. In this way Mach brings his readers very close to the actual process of scientific investigation.

6. Laws are hypotheses

If the term 'law' means 'an experimental generalisation' then the expression of a 'fact' such as:

> 'At 0°C and 76 cm Hg pressure, 40·0 g of lead nitrate saturates 100 g water'

is also a law. For convenience, let this law be referred to as L.

This looks like a categorical statement, but it is hypothetical. This must be so because the reference is to *all* lead nitrate, *all* water, and to the *future*. We have no experience of *all*, and no experience of the *future*. L must therefore be a hypothesis, representing our *expectation* rather than our *completely certain knowledge*. We are obliged to agree with Karl Popper, that "Laws . . . are always hypotheses."[1]

Since L is not established as categorical, and will always remain hypothetical however much experience there may be, there is strictly no problem of induction at all.[2] We cannot draw categorical all-statements out of our experiences. Logically, our lives are conducted along countless question marks and innumerable hypotheses. The various hypotheses of science differ in degree but not in kind. L, our example of the solubility of lead nitrate, is a hypothesis just as the kinetic theory of gases

[8] R. Deakin, *Euclid: Books I–IV*, pp. 14, 15.

[1] K. R. Popper, *The Logic of Scientific Discovery*, London: 1959, p. 247.

[2] Ibid., p. 40.

is a hypothesis. As Popper puts it, "scientific theories are universal statements".[3]

Aristotle is correct in thinking that knowledge is necessarily of the universal. That species of knowledge which we now call science is a system of theories, all of which are hypothetical.

The empirical sciences are systems of theories. The logic of scientific knowledge can therefore be described as a theory of theories.[4]

Popper explicitly opposes the view that science is reducible to sense-perceptions:

The doctrine that the empirical sciences are reducible to sense-perceptions, and thus to our experiences, is one which many accept as obvious beyond all question. However, this doctrine stands or falls with inductive logic and is here rejected along with it. I do not wish to deny that there is a grain of truth in the view that mathematics and logic are based on thinking, and the factual sciences on sense-perceptions.[5]

Although Popper admits that science has an empirical basis, he uses the disparaging metaphor of a swamp to represent this empirical basis, and he contrasts the swamp with the piles let down from the intellect of man into his confused and vaguely characterised experience:

The empirical basis of objective science has thus nothing 'absolute' about it. Science does not rest upon rock-bottom. The bold structure of its theories rises, as it were, above a swamp. It is like a building erected on piles. The piles are driven down from above into the swamp, but not down to any natural or 'given' base; and when we cease our attempts to drive our piles into a deeper layer, it is not because we have reached firm ground. We simply stop when we are satisfied that they are firm enough to carry the structure, at least for the time being.[6]

Theory is secure and strong, the intellectual parallel bars on which the proper game of logical deductive reasoning can be played. Although one must remember that the premises and conclusions of the logical processes are all hypothetical.

Mach might have objected, to Popper, that it is unfair to represent something that is not 'there' at all by the metaphor

[3] Ibid., p. 59. [4] Ibid. [5] Ibid., p. 93. [6] Ibid., p. 111.

of strong steel girders. When Mach says that "Die Natur ist nur einmal da",[7] undoubtedly he means that the web of the sensuous elements is 'there'. Mach is not true to his antimetaphysical creed in saying that anything at all 'is there'. But I maintain that metaphysical questions cannot be shelved indefinitely. If anything is 'there' in the natural world, it is Popper's swamp, not the piles.

Possibly Mach would have preferred to turn Popper's metaphor around, representing the experience of man by steel girders (L *looks* very strong) and theories and conceptual systems by swamp. It is very characteristic of theories that, like swamps, they can readily be changed. Nevertheless it is easy to exaggerate the difference between Mach and the best modern critics of science. I conclude with a passage, quoted in part previously, which should always be remembered when Mach's contribution is being judged:

> But for scientific purposes our mental representations of the facts of sensual experience must be submitted to *conceptual* formulation. . . . This formulation is effected by isolating and emphasising what is deemed of importance, by neglecting what is subsidiary, by *abstracting*, by idealizing. . . . Without some preconceived opinion the experiment is impossible, because its form is determined by the opinion. For how and on what could we experiment if we did not previously have some suspicion of what we were about? . . . The experiment confirms, modifies, or overthrows our suspicion.[8]

7. "The analytic functioning of empirical laws"[1]

'The boiling-point of ethanol at standard pressure is 78°C' is a law, in the same way that L, of the previous section, is a law. According to Arthur Pap, it would be more correct to say that 'The boiling-point of ethanol at standard pressure is 78°C' *was* a law. If the student in 1971 suspects a liquid to be alcohol, he measures the boiling-point and takes 78°C as a proof that the liquid is in fact ethanol. This means that 'boiling at 78°C' has *become the definition* of alcohol, or part of the definition of alcohol. The suggestion that the student's measurement should be taken

[7] PP., p. 230, and P., p. 199. [8] M., p. 161.

[1] A. Pap, *The A Priori in Physical Theory*, New York: 1946, p. 28.

as further evidence for the law stated above would be dismissed at once as ludicrous. This has been usefully described as "the analytic functioning of empirical laws":

> Suppose, now, that, having verified Hooke's law by successive additions of weights to the test wire, we find that upon application of further weights we obtain points deviating from the straight line farther than is justifiable by the assumption of a certain interval of accidental error. In all likelihood, this discrepancy will not be interpreted, by the experimental physicist, as a refutation of Hooke's law, but as an indication that the elastic limit of the wire has been exceeded. Inasmuch as he argues "stress is not proportional to strain, therefore the elastic limit has been exceeded", he is employing Hooke's law as a definitional criterion, since from the falsity of the consequent he infers that the condition which delimits the applicability of Hooke's law is not satisfied.[2]

A good example of the same logical point is found in the law of constant composition, one of the first principles of chemical science. If, following Boyle, the compound is defined as an association of elements A, B . . . such that

A + B . . . make Z,

where Z stands for a new substance whose properties bear no simple relationship to the properties of A, B . . ., then it is logically proper and correct for the chemists to go on to *discover* the law of constant composition as an empirical generalisation. If however the compound Z is *defined* as that which has a constant gravimetric composition, then the law functions as a definition or, in Pap's language, becomes analytical. As a teacher of chemistry, I have for many years advocated the desirability[3] of maintaining Boyle's definition and so preserving the status of the constant composition of compounds as a law. Provided *one can see in the compound something more than a constant composition,* then there is a place for the empirical law concerning the compound's composition. Pap makes the same point for the test wire and Hooke's law. Such a wire more or less 'obeys' Hooke's law: but *this does not exhaust its qualities.*

[2] Ibid., p. 30.

[3] J. Bradley, 'The Classical Molecular and Atomic Theory of Chemistry,' *School Science Review*, **39**, no. 137, 1957, p. 12.

It has, no doubt, various metallurgical characteristics, it is made of some specific metal and so on. There could therefore be a place for Hooke's law after all.

The idea of the perfect object is an extrapolation of Arthur Pap's theme. The general rules according to which the perfect object operates, can hardly be called laws of nature if the object is a creation of the human mind. In so far as the lever is a self-created ideal object, the so-called law of the lever is part of the definition of the ideal object.

The tendency of Pap's argument is to diminish the number of laws by reducing some of them to definitions. The tendency of Popper's discussion is to make the term 'law' redundant, for laws are hypotheses and so are other 'all-statements'.

8. Explanation

The general tenor of Mach's account of the theoretical element in science is that it is better to *describe* than to *explain*, and that a *theory* is correctly to be understood as a mode of *indirect description*. Nevertheless he frequently uses the term 'explanation' (Erklärung) and in one passage he states fairly clearly what he means by it:

> When a new mathematical or scientific fact is *demonstrated*, or *explained*, such demonstration also rests simply upon showing the connexion of the new fact with the facts already known; . . .[1]

H. Dingle takes much the same view:

> I think it is generally acknowledged now that, in science, explaining a phenomenon simply means stating its relations with other phenomena.[2]

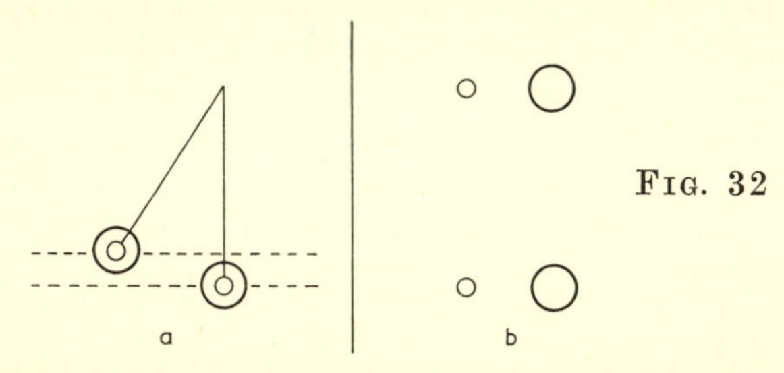

FIG. 32

[1] P., p. 362. [2] H. Dingle, *Through Science to Philosophy*, p. 26.

As an example of such explanation we may consider the mechanical systems depicted in Fig. 32. Fig. 32a shows two simple pendulums, the one behind the other, whose bobs are of very different masses. They keep time together. Fig. 32b shows two very different masses falling freely; they also keep together. If we *see the pendulums as falling weights*, we feel that the pendulum phenomenon has been, to some extent, 'explained' because it has been 'related' to another phenomenon.[3] Explanation in this simple sense is represented in the next diagram,

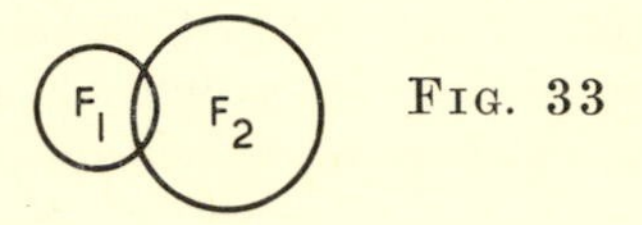

FIG. 33

Fig. 33. The two areas of 'fact' or 'law', F_1 F_2, overlap; in the example of Fig. 32, both systems exhibit falling weights.

9. Explanation by analogy of the first kind: fact—fact analogy

The example of Fig. 32 is offered as an *explanation*; it is also an *analogy*. If to explain is to connect two orders of fact, the explanation is an analogy; in other words, analogy must have a tendency to explain.

Resemblance is partial identity. The characteristics of similar objects agree in part, differ in part. *Analogy* is nevertheless a *special* case of *resemblance*.[1]

A much more important kind of analogy arises from what Mach calls "the agreement in logical relations"[2] between two orders of fact and which constitutes "an effective means of mastering heterogeneous fields of facts in unitary comprehension".[3] The classical example is Ohm's law,

$$\text{current} = \frac{\text{pressure difference}}{\text{resistance}},$$

which holds for currents of water, electricity and heat. The force

[3] I cannot recollect the source of this example which is not original.
[1] E., p. 220. [2] P., p. 250. [3] Ibid.

of this important analogy does not lie in any resemblance between water, and heat, or even between the effects of these natural agents. For example, water has mass but heat has no mass; a flow of electricity sets up a magnetic field whereas neither the water nor the thermal current does this. The force of the analogy lies rather in the *common* relationship or simple proportionality between current *in general* and potential gradient *in general*. Such an analogy can be an inspiration to the investigator: "Ohm forms his conception of the *electric* current in imitation of Fourier's [*heat* current]."[4] Indeed, this is an understatement. Ohm derived his famous 'law', which is in part a definition of electrical potential, from Fourier's work. As P. G. Tait said, "the mere application of one of its simplest portions to the conduction of electricity has made the name of Ohm famous."[5]

Fig. 33 will not serve to depict the Fourier–Ohm analogy. In Fig. 34, three areas of fact—areas concerning the flow of water,

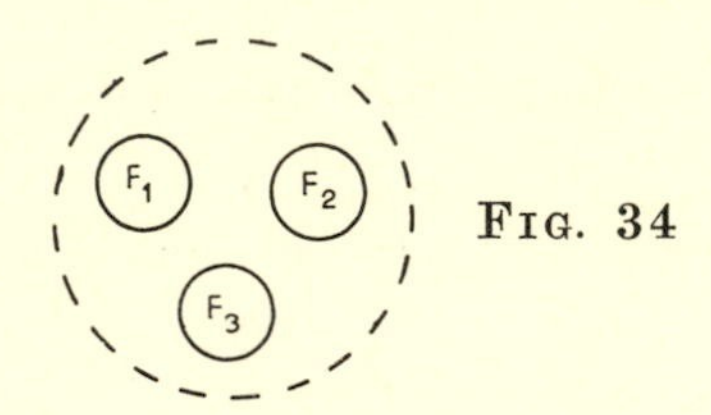

FIG. 34

electricity and heat—are represented by the circles F_1 F_2 F_3. The dotted circle which encloses F_1 F_2 F_3 is what Mach calls "the agreement in logical relations." It amounts to an algebraic form which comprehends three orders of phenomenon. The resemblance which makes the analogy is an abstract one.[6] There is no logical nexus between any two of the terms F_1 F_2 F_3. So:

> Conclusions based on resemblance and analogy are strictly no affair of logic, at least not of formal logic; rather are they merely a *psychological* matter.[7]

Ohm's law for electricity cannot be *deduced* from Fourier's

[4] P., p. 249.
[5] T. Preston, *Theory of Heat*, London: 1929, p. 596, quotation from P. G. Tait. [6] E., p. 220. [7] E., p. 225.

conductivity equations; Ohm can however be *inspired* by Fourier.

Two related examples of analogy of this kind (Fig. 34) are given by Mach. The former was first drawn by Zeuner, the latter is Mach's original conception. They are given here with minor adaptations;[8] they show Mach at his best and suggest to us that physics, like religion, is well studied *comparatively*.

Suppose that in the perfectly reversible heat engine the heat taken from the source at temperature T_1 is $Q + Q'$, and the heat given up to the sink at T_2 is Q. Q' represents then that part of the heat which is transformed into mechanical work. For convenience, Zeuner used the ordinary Carnot equation

$$\frac{Q + Q'}{T_1} = \frac{Q}{T_2}$$

in the equivalent form

$$-\frac{Q'}{T_1} + Q\left(\frac{1}{T_2} - \frac{1}{T_1}\right) = 0. \tag{1}$$

Fig. 35 represents a capacious or infinite reservoir of water standing at height h_1, this height h_1 being the analogue of

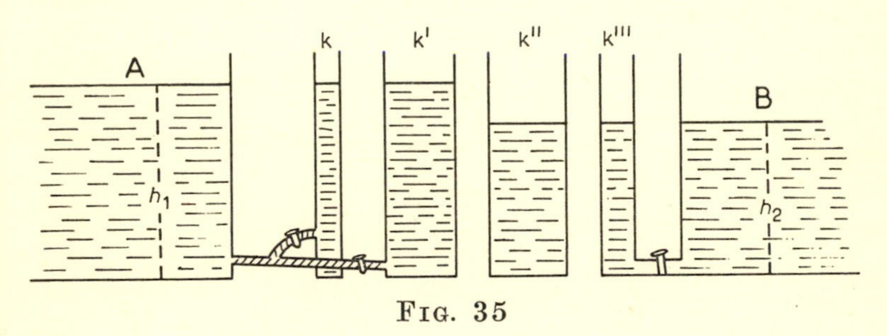

Fig. 35

temperature T_1. The letters k, k′, k″, k‴ represent *one and the same vessel*, a vessel whose sides can be made to expand further apart or contract nearer together as required. The familiar stages of the Carnot cycle are reproduced as follows.

i. The vessel k, when it is most narrow, is in contact with A, and the water stands at level h_1 in k.

[8] W., pp. 328–33,

ii. The sides of k are expanded to k′ and throughout this change k k′ remains in communication with A. A weight P (physical dimensions MLT^{-2}, Mach's choice of the letter P is not a happy one) of water is taken in from A during the expansion. The level in k′ is h_1, and the level in A remains h_1.

iii. The vessel is now expanded further, when cut off from the reservoir A. k′ thus becomes k″ and the level h_1 falls to h_2. The level h_2 is also the level or height of the water in the infinite reservoir B, which corresponds to the sink of the heat engine.

iv. The k vessel is now put into communication with B. It is contracted to k‴. Its dimensions are such that the weight P of water is discharged to the vessel B. The level remains h_2 in k‴, and the level in B also remains h_2. k, k‴ represent the vessel with an equal water content; k′, k″ represent the vessel again with an equal, but greater, water content.

v. The sides are further contracted, the k vessel being cut off from both reservoirs. When the level is again h_1, the whole cycle can be repeated, and so on.

In pointing the analogy,[9]

$$h_1 \equiv T_1$$
$$h_2 \equiv T_2,$$

but it is somewhat more difficult to see what Q' and Q stand for. The difficulty is resolved by thinking of Q' first. In the transition k′ → k″ (stage iii), a weight P of water changes its mean height by the quantity

$$\frac{h_1 - h_2}{2}.$$

There is therefore, per cycle, a loss of potential energy equal to

$$\frac{P}{2}(h_1 - h_2).$$

This evidently could be turned into external work. Zeuner would therefore write

$$\frac{P}{2}(h_1 - h_2) \equiv Q',$$

[9] The symbol $\equiv$ here means 'is analogous to'.

and by formal analogy with the usual Carnot relationships, we are obliged to add

$$\frac{P}{2} \cdot h_2 \equiv Q.$$

The whole analogy is now checked by substituting each term of the left-hand side of eqaution (1) by the analogical term. This gives:

$$-\frac{P}{2} \cdot \left(\frac{h_1 - h_2}{h_1}\right) + \frac{P \cdot h_2}{2} \cdot \left(\frac{1}{h_2} - \frac{1}{h_1}\right)$$

or

$$-\frac{P}{2} + \frac{P \cdot h_2}{2h_1} + \frac{P}{2} - \frac{P \cdot h_2}{2h_1}$$

which is 0, as it ought to be. The reader should note that the expressions $(P/2) \cdot h_1$ and $(P/2) \cdot h_2$ are evidently correct for potential energy lost by the source and gained by the sink; however capacious A and B, the *mean* heights of the water are $h_1/2$ and $h_2/2$.

From 1876 to 1896 Mach was accustomed to use a similar analogy in teaching electrostatics. Fig. 36, not given by Mach,

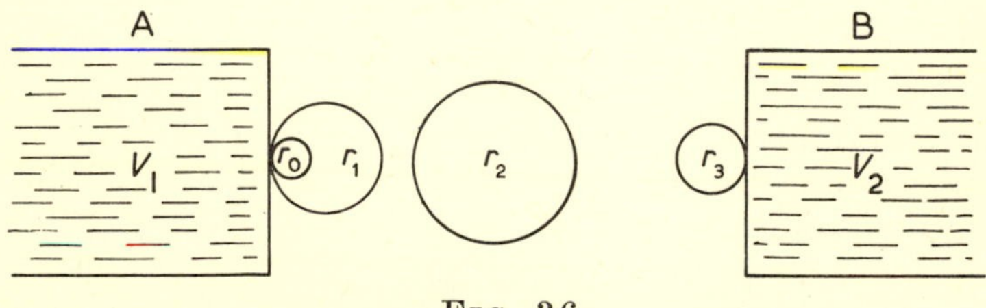

FIG. 36

shows two immense or infinite bodies A and B at electrical potentials V_1 and V_2 respectively. Corresponding to the adjustable cylindrical water vessel k, this time we have an adjustable metal spherical ball. It is assumed that $V_1 > V_2$. Corresponding to the five stages of Zeuner's analogy, we have the following:

i. The metal ball, at its smallest with radius r_0, is in contact with A, and the electricity on the ball is at the potential V_1.

ii. The ball is expanded to radius r_1, and during the expansion remains in contact with A. As a result of the increased electrical capacity, the ball takes in electricity P from A, at the pressure V_1.

iii. The ball is cut off from A, and is further expanded to the radius and capacity r_2. The value of r_2 is such that by this further expansion, the potential falls to V_2. During the expansion $r_1 \rightarrow r_2$, there is no loss or gain of electricity by the ball. Therefore,

$$r_1 V_1 = r_2 V_2.$$

iv. The metal ball is now put into electrical communication with B. As the radius shrinks to r_3, the electricity P is given to B at potential V_2, and the potential of the ball also becomes V_2.

v. Finally the ball is insulated, made to shrink back to radius r_0, and its potential is again V_1.

Again there is no loss or gain of electricity, and so

$$r_3 V_2 = r_0 V_1.$$

The cycle can then be repeated. The analogy is again pointed in the same way. We can begin, as before, with

$$V_1 \equiv T_1$$

and

$$V_2 \equiv T_2.$$

The energy given to the ball from A is $\frac{1}{2} \cdot V_1^2(r_1 - r_0)$. The energy returned to B is $\frac{1}{2} \cdot V_2^2(r_2 - r_3)$. The difference can be turned into work, so that

$$\tfrac{1}{2} \cdot V_1^2(r_1 - r_0) - \tfrac{1}{2} \cdot V_2^2(r_2 - r_3) \equiv Q'.$$

Further,

$$\tfrac{1}{2} \cdot V_2^2(r_2 - r_3) \equiv Q.$$

The expression corresponding to the left side of equation (1) is therefore

$$-\frac{1}{2} \cdot V_1(r_1 - r_0) + \frac{1}{2} \cdot \frac{V_2^2}{V_1}(r_2 - r_3) + \frac{1}{2} \cdot V_2^2(r_2 - r_3)\left(\frac{1}{V_2} - \frac{1}{V_1}\right),$$

which reduces to

$$-\tfrac{1}{2} \cdot V_1(r_1 - r_0) + \tfrac{1}{2} \cdot V_2(r_2 - r_3);$$

from the additional information that

$$r_3V_2 = r_0V_1$$

and

$$r_1V_1 = r_2V_2,$$

this equals 0, as it should.

Mach summarises the three cases by a formal statement that the generalised form of equation (1), i.e.

$$-\frac{W'}{V_1} + W\left(\frac{1}{V_2} - \frac{1}{V_1}\right) = 0 \qquad (2)$$

holds for all. In equation (2),

W' = work done,
W = energy 'lost to the lower potential V_2',
V_1 = the higher potential and
V_2 = the lower potential.

It is necessary in developing this kind of analogy to emphasise, as Mach does, that "analogy is not identity".[10] From Fig. 34, we may affirm that F_1 is expressed by A, where A is an algebraic form, and also that F_2 is expressed by A, viz. the same mathematical form. This is like the premises of a syllogism with undistributed middle term; no conclusion follows. Mach in fact adds a cautionary analysis of the *negative* analogy between heat and electricity. The reader is referred to the *Principles of Heat* for the details.[11]

10. Explanation by analogy of the second kind: fact—model analogy

Mach's treatment of the molecular-atomic hypothesis is enlightened and subtle. The molecules and atoms of chemical and physical theory are conveniently called a model. This model is taken by Mach as *one side* of an analogy. This bald statement of Mach's view may clarify what follows. Fig. 37 should be compared with Fig. 34. As before, the dotted circle represents "the agreement in logical relations", and F_1 F_2 stand for areas of

[10] W., p. 334. [11] W., pp. 334–6.

fact. But M is an intellectual construction and not an area of fact. The resemblance which constitutes the analogy is abstract

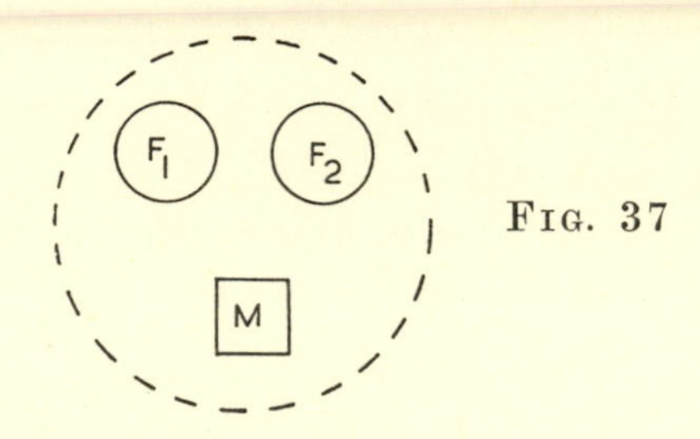

FIG. 37

as before; but it is no longer between areas of fact, it is between one or more areas of fact F and the so-called model M.

(a) *Models*

Towards the end of the *Mechanics* Mach explicitly describes the atomic theory as a model:

> The atomic theory plays a part in physics similar to that of certain auxiliary concepts in mathematics; it is a mathematical *model* for facilitating the mental reproduction of facts.[1]

Modern writers use the term 'model' in almost the same way.[2] To Mach, the atomic theory is the term M of the second kind of analogy (Fig. 37). It is not necessary here to determine how many kinds of theory there may be, but it is useful to distinguish between such a theory as the atomic-molecular theory, called by Mach and others a model, and the very abstract kind of theory represented by, say, the general theory of relativity. In the important passage now quoted Mach refers to the atomic theory as an analogy:

> And when one adds that this view [i.e. the atomic theory] has also led to new discoveries, whilst the analogy has retained its validity beyond the expectation at the time of its discovery, one cannot be surprised at the esteem it enjoys among chemists. But that cannot prevent us from laying bare the theoretical kernel, as indeed we have done, and from viewing the accidental superficial accretions [additions] to the theory as images which need not be taken seriously. Of any new theory of chemical phenomena we must surely require

[1] M., p. 589. [2] R. B. Braithwaite, *Scientific Explanation*, p. 93.

that it shall contribute at least as much as the more or less abandoned atomic theory which it replaces.[3]

(b) *Molecule and table*

For Mach, a 'body' or a 'thing' is not presented to our sense-perception; the body and the thing are convenient mental symbols for groups of elementary sense-perceptions. A 'substance' such as chlorine, again, is not presented to our sense-perception; the term chlorine refers to a closely cohering group of sense-perceptions or, more accurately, possible sense-perceptions. Now Mach considers that the molecules and atoms are 'things' in exactly the same way, and that they too must be mental symbols:

If ordinary "matter" must be regarded merely as a highly natural, unconsciously constructed mental symbol for a relatively stable complex of sensational elements, much more must this be the case with the artificial hypothetical atoms and molecules of physics and chemistry. The value of these implements for their special, limited purposes is not one whit destroyed. As before, they remain economical ways of symbolising experience. But we have as little right to expect from them, as from the symbols of algebra, more than we have put into them, and certainly not more enlightenment and revelation than from experience itself.[4]

Mach is as sure as Hume that only we men can feel; there is nothing more real, nothing more actual (wirklich) than our *feelings*. It would therefore be absurd to ask how a complex molecular biochemical system, which is a mental symbol for a man, *feels* or *thinks*. The following words are perhaps the most eloquent in all Mach's writings:

What those ideas are with which we shall comprehend the world when the closed circuit of physical and psychological facts shall lie complete before us, (that circuit of which we now see only two disjoined parts), cannot be foreseen at the outset of the work. The men will be found who will see what is right and will have the courage, instead of wandering in the intricate paths of logical and historical accident, to enter on the straight ways to the heights from which the mighty stream of facts can be surveyed. Whether the

[3] W., pp. 358–9. [4] A., p. 311.

notion which we now call *matter* will continue to have a scientific significance beyond the crude purposes of common life, we do not know. But we certainly shall wonder how colours and tones which were such innermost parts of us could suddenly get lost in our physical world of atoms; how we could be suddenly surprised that something which outside us simply clicked and beat, in our heads should make light and music; and how we could ask whether matter can *feel*, that is to say, whether a mental symbol for a group of sensations can feel?[5]

It is agreed that atoms and molecules are 'little things'. Mach, in some degree, objects to these 'little things', but he objects to them not because they are 'little' but because they are 'things'. He has no time for the *Ding an sich*, for the primary Cartesian qualities taken as objective but unperceived reality, nor for the real atoms and void of Democritus. If, however, atomism is simply a tool for the economical expression of experienced phenomena, and in this form Mach takes no objection to it, then it could be at last dispensable. Even those who could not possibly be brought to regard the matter quite in this light must agree with Mach in respect to the order of priority between "the living stream" of experience and "the mosaic play with stones", more prosaically described as the working out of the implications of a conceptual system. Mach consistently places facts (die Tatsachen) at the centre of the activity called science. As a teacher of chemistry to children, I have noticed how such hard 'stones' as

$$3Cu + 8HNO_3 = 3Cu(NO_3)_2 + 4H_2O + 2NO,$$

which even as atomic mythology have always been arid and have now become heretical (1971), may turn their minds away from chemistry. The children come to school for chemical experience; they ask for bread. "If a son shall ask bread of any of you that is a father, will he give him a stone?"[6]

It has often been said that to the late Lord Rutherford the α-particle was as real as a table or a chair. The same could be said for Mach. To him the table, the benzene molecule, and the hydrogen atom enjoy the same status. To affirm the reality or

[5] P., pp. 212, 213. [6] St. Luke, 11, 11.

otherwise of such 'things' is an entirely unnecessary excursion into metaphysics which is not required in the effective practice of physical science.

(c) *Kinetic molecular model*

In his first important publication,[7] Mach gives a most interesting discussion of the kinetic molecular theory. Although to physicists like Clerk Maxwell or Boltzmann it might have appeared unlikely that the molecules could or would ever be seen or touched, yet when they set about building their model, they invested the molecules with the kind of properties which would render them in principle visible and tangible if our sense-perceptions of sight and touch were to be vastly extended and improved. The molecules were little elastic balls dashing about in three-dimensioned space, and exerting a pressure on the walls of the containing vessel. Clerk Maxwell's molecules were a model, in the design of which certain limitations of the tangible and visible were imposed. Yet such a model only selectively and partially resembled the macro-world of common life. It was not considered necessary for the molecules to sing or to be coloured, although as Mach says:

> There is no more necessity to think of what is merely a product of thought spatially, that is to say, with the relations of the visible and tangible, than there is to think of these things in a definite position in the scale of tones.[8]

We can therefore never truly derive from such a model any valid interpretation of a colour, of a musical note, or of a feeling of warmth or temperature. In building the model there were negative as well as positive conceptual decisions. There were no 'warmth' and no 'colour' ingredients in the outfit from which the model was constructed. To produce such out of the model later would be a conjuring trick rather than physics. This is precisely the point made by the late Max Born in the passage quoted earlier;[9] temperature is "a foreign element" in the kinetic theory of gases for it is excluded from the start.

How then shall a man prove Boyle's law? Will he go to his bench, measure the various pressures and volumes of a gas and

[7] G., 1872. [8] G., p. 51. [9] M. Born, *Atomic Physics*, p. 7.

make an inductive leap to the generalisation? Or will he derive it by logical reasoning from the axioms or premises of the kinetic theory? As Bridgman would have expressed it, there are two Boyle's laws and they must be distinguished as:

Boyle's law$_1$—an experimental generalisation;
and Boyle's law$_2$—an item of a conceptual system.

Strictly, the proof of Boyle's law$_2$ in no way strengthens the 'proof' of Boyle's law$_1$. The second use of the term 'proof' in the last sentence is based on the logic of Mill and is rejected by Mach and Popper. Referring back to Fig. 37, it could be said that Boyle's law$_1$ is—not quite accurately for I do not forget that empirical laws are hypothetical—a term of the type F_1; Boyle's law$_2$ is a term of the type M. Again from the premises,

F_1 has an algebraic form A
and M has the same algebraic form A,

nothing follows. The middle term is undistributed. For Mach, the kinetic theory is one side of an analogy of a special kind—of the kind which I refer to here as the second kind.

At least in one place, Mach seems to regard the exclusion of temperature, colour and sound from the particles of the kinetic theory as a weakness in that theory.[10] In this he is wrong. We can have our models as we like. We could incorporate the assumption that an individual molecule has a degree of temperature, in addition to having motion and size. This would be tantamount to abandoning the kinetic theory in its present form, and replacing it by another theory. The question of the temperature of a molecule would be reopened. It is interesting to find such close agreement with Mach on the general question of molecules and atoms in a work published as recently as 1963:

> Nevertheless, the layman would do well to note in the meantime that as the terms "molecule", "atom", "electron", and others denoting microcosmic particles are used in the context of physical theories, they cannot be meaningfully combined with predicates designating temperatures or colors—or even smells! . . . For whereas a man who affirmed the existence of tiny pebbles that are

[10] P., pp. 104, 105.

colorless and devoid of temperature could justly be accused of talking nonsense, the analogous affirmation of the existence of colorless and temperatureless atoms will appear nonsensical only if one confuses the theoretical language of the scientist with the "thing-language" of everyday life. . . .[11]

Mach would not have cared for Pap's rather careless use of the term 'existence'.

Such is Mach's position with respect to the kinetic molecular theory. It is a safe position. The kinetic theory, and other theories, are worthy of respect for they facilitate the work of economic communication and are helpful in teaching. Undoubtedly Mach makes too little of the fertilising heuristic power of theories. If the kinetic theory is merely the conceptual side of an analogy, its function as the channel through which the very spirit of discovery flows becomes hard to account for.

(d) *Mach and chemistry*

In both chemical and physical changes there is a change of some kind to a material substance, but throughout both there is no total mass change. What then is the difference? In the physical change, e.g. the elastic extension of a wire, "a single property"[12] changes, whereas in the chemical change "the entire complex"[13] disappears and is replaced by a new one or some new ones. Mach is quite correct in finding the disappearance of one or more substances, and the appearance of one or more new substances, at the centre of any philosophy of chemistry. If a compound disappears, or if a compound appears, then a chemical change is said to occur.

The body of the science of chemistry is made up of "*general* phenomenological laws",[14] or—perhaps more clearly—of a wealth of facts about the chemical compounds and the correlative chemical changes. If this wealth of facts be divided into qualitative and quantitative, the former class is not less important than the latter. If the chemist cannot rest here, but must develop the formulae and equations which set forth a kind of atomic model of the phenomena, he must not fool himself

[11] A. Pap, *An Introduction to the Philosophy of Science*, London: 1963, pp. 55, 56. [12] W., p. 355. [13] Ibid. [14] W., p. 356.

into regarding this atomic model as explanatory.[15] Mach's dislike of the idea of explanation amounts to a prejudice. But he is correct in thinking that the explanatory value of the atomic model is exaggerated by students of chemistry. When they write the 'equation'

$$2Na + Cl_2 = 2NaCl$$

they think they are explaining the burning of sodium in chlorine. This is hardly the case, and Mach is correct. It is evidently true that atomic interpretations are often presented to beginners at such a time and in such a way that they can have little or no explanatory value. Nevertheless, it is a pity that Mach omits to note the Kantian interpretation of the chemical equation. This has been elaborated earlier, in the section on Kant. What is permanent or persistent, what remains 'equal', is the particular value of a 'non-O' concept, something beyond the reach of sense-perception. Kant's philosophy requires us to find something persistent. We can achieve this without atoms, but not with sense-perceptions or with concepts that bear the 'O' relationship to sense-perceptions. Mach should have considered the possibility of replacing the *Ding an sich* by the 'non-O' concept.

11. Positivism

Comte's *Cours de Philosophie Positive* was completed in 1842. At the outset of the enormous treatise he describes how the human mind tends to grow through three states or stages characterised as

I Theological
II Metaphysical
and III Positive

in that order. At stage I the mind seeks out the ultimate nature of things, looks for first causes, and attributes events to the agency of supernatural powers. Having grown out of this, the mind enters the second phase. The gods are replaced by abstract

[15] Ibid.

forces, and the superstitions of religion by the credulities of moral philosophy. At last, in what Comte calls the positive stage, the mind has learnt that it cannot know much, and that physical science, understood in the manner of Hume rather than Kant, supplies the model or type of any true kind of knowledge.

Positivism is a temper of mind and it is characterised by Comte and his successors in negative terms. It is the determination of the mind to have *nothing* to do with metaphysics, to *decline* to look for realities hidden behind appearances, and to *refrain* from explanation in physics. As Meyerson neatly expresses it, "Positivism means . . . a complete abstention from metaphysics".[1] It is strange, as James Martineau said in an essay on Comte, that something so negative should be called positivism.

There is only one important reference to Comte in the whole of Mach's writings. It is unexpected and uncharacteristic, yet in it Mach expresses his general agreement with Comte's notion of the three stages:

This play of the imagination around what is experienced or seen, this *poetry* of life, is the first elevation rising above the everyday, above the breathless bearing of the burden of life. Although this poetry, if translated uncritically into practical terms, may often bear bad fruit, as we have just seen, yet it is nonetheless the beginning of mental development. When these phantasies are brought into relationship with sensory experience in the serious intention of illuminating the latter and, on the other hand, of allowing themselves to be rectified by it, then emerge, stage by stage, religious, philosophical and scientific ideas (A. Comte). Let us therefore consider this poetic imagination which completes and modifies everything experienced.[2]

When Comte calls his third stage 'positivism', he means that to him theology and Kantian philosophy are just nothing at all. Mach sees them rather as activities in which the element of phantasy plays a dominant role. Moreover he admits that they are necessary to complete physics itself, although the main part

[1] E. Meyerson, *De l'Explication dans les Sciences*, pp. 19, 20.
[2] E., p. 99.

of physics is what is experienced through sense-perception. Mach is far too good a physicist, and far too candid a commentator, to reduce the intellectual-artistic ingredient of physical science to the scale required by Comte's ideas. Nevertheless Mach, like Comte, is generally opposed to introducing metal physical questions or he at least frequently professes to be; like Comte, he cannot tolerate the notion of a reality behind appearances.

Nowhere in his main writings does Mach call himself a positivist, although in the posthumously published *Leading Thoughts*, he tacitly accepts the title as conferred upon him by Max Planck:

> Now I am able to say that P. [Planck] . . . does not judge my positivism rightly when he sees it as a reaction from the failures of atomistic speculations. Were the kinetic physical world-picture, which it is true I take to be hypothetical without thereby wishing to discredit it, capable of "explaining" all physical phenomena, I should still maintain that the manifold diversity of nature is not exhausted. To me *matter*, *time*, and *space* are also *problems*, towards the solution of which, incidentally, physicists like Lorentz, Einstein and Minkowski gradually approach. Nor is physics the whole world; there is *biology* too, and it should have an essential place in the world picture.[3]

In these words, written when he was over seventy, Mach shows himself to be a moderate, wise and catholic thinker. He accepts the title 'positivist' but objects to the narrow interpretation of this as merely an abstention from atomic theorising. Indeed he has no wish to discredit the molecular kinetic picture, although he makes it clear that such a picture is theoretical or hypothetical. He appreciates the advances in the new relativity physics of Einstein and others. Finally he reminds us that 'All is physics' is no less a heresy than Thales' doctrine, 'All is water'. Mach believed that the contribution of Darwin to science is not less significant than the contribution of Newton, and this is the case.

What, we may inquire, is positive about Mach's form of positivism? It is that science is no more than the direct or

[3] E. Mach, *Die Leitgedanken*, Leipzig: 1919, p. 15. This little book contains two magazine articles both written in 1910, and published posthumously.

indirect description of natural phenomena and facts. Inquiry into what reality lies behind what appears, into cause and necessary effect and into explanation understood in the old-fashioned sense, is not legitimate.

. . . what is called a *theory* or a *theoretical idea,* falls under the category of what is here termed indirect description. . . .
Does description accomplish all that the inquirer can ask? In my opinion, it does.[4]

12. Description

Mach's final view of science is that its basis is a rich sensuous experience; it is better to feel than to think, better to describe than to explain. Certainly the feeling and the describing are prior to the thinking and explaining. Like the maid in Wordsworth's poem,[1] Mach welcomes what is given and craves no more. He intensely loves what he sees, and wishes for nothing better. The few nooks to which his happy feet are limited are sufficiently exquisitely framed. He feels and observes, and would refrain from making theories. In the end he becomes an artist; but he is a sensitive being *before* he is a creative soul.

But the feelings and observations must be communicated and this requires description in words; for science is a kind of public property, not a private experience. Mach's division of description into two kinds, the direct and indirect, is valuable. Briefly, his teaching can be epitomised in this way:

Direct Description ≡ The Laws of Science;
Indirect Description ≡ The Theories of Science.

Karl Pearson also refers to scientific law as a species of description:

Scientific law . . . is a brief description in mental shorthand of as wide a range as possible of the sequences of our sense-impressions. . . . in the present work we have learnt to look upon all science as a *description,* . . .[2]

[4] P., pp. 240, 241, 252, 253.
[1] This paragraph is a paraphrase of Wordsworth's *Prelude,* Book XII, ll. 174–207.
[2] K. Pearson, *The Grammar of Science,* pp. 98, 320.

Mach clearly connects the ideas of law and direct description in the *Popular Lectures*:

> Again, to save the labour of instruction and of acquisition, concise, *abridged description* is sought. *This is really all that natural laws are.*[3]

When 'theories of science' are equated with 'indirect description', the term 'theory' has a wide connotation. Hypothesis, fact-fact analogy, fact-model analogy, thought-experiment and perfect object, and abstract mathematical analysis all, in various ways, contribute to the work of indirect description. It might be thought difficult or even impossible to maintain that such a formula as

$$G\frac{m_1 m_2}{r^2}$$

for the force between two mass points m_1, m_2 at distance r is any kind of description of something represented to sense-perception. In one of his last untranslated *Popular Lectures*, Mach makes it quite clear that he does nevertheless regard this formula as an example of indirect description.[4] Against Mach's opinion, mass is completely or absolutely out of the range of any sense-perception at all; we have no experience of bodies whose masses are points; we do not feel the force between the sun and a planet, and so on. On the other hand, in favour of Mach, it is true that such a formula represents also a terminus of the human reason; in a way, it has as little to do with our minds as it has with our organs of sense-perception. Why—and Newton declined to ask the question—should a mass m_1 attract another mass m_2 at a distance r from it? There is no answer in classical dynamics; all we can do is to describe the 'brute fact'. The formula is the clever way which Newton found of doing this. When Mach calls Kepler's first law a direct description and Newton's formula an indirect description, his legitimate purpose is to remind us that science is based on the 'given', not the 'rational'. There is here, I believe, a common ground shared by religion and science; devotees of both must initially accept what they find to be 'there'. Theological systematisation and

[3] P., p. 193. [4] PP., p. 426.

scientific theory alike are secondary and later phases of the respective disciplines. Evidently Newton's formula is *not less than* an abbreviated description of what *is*, or of what *was* when Brahe made his observations; if the formula were

$$G\frac{m_1m_2}{r^3}$$

then the planets would not follow elliptical orbits. However well established as a part of a theoretical system, the formula is at once rejected if it indirectly describes what is not the case. In exactly the same way the famous hexagon formula for the molecule of benzene is an indirect description of whole ranges of experienced fact; there is no legitimate octagon alternative. Chemistry, like physics and also like religion, is not logical or mathematical in the first place.

One can see without difficulty that what we call a *theory*, a *theoretical* idea, or a supplement to a theory, fall into the category of the indirect description.[5]

Understood in this way Mach's descriptionism is true and extremely important. But he carries it too far:

It would appear . . . not only advisable, but even necessary, with all due recognition of the helpfulness of theoretic ideas in research, yet gradually, as the new facts grow familiar, to substitute for indirect description *direct* description, which contains nothing that is unessential and restricts itself absolutely to the abstract apprehension of facts.[6]

This perhaps could be taken as a general statement of Hooke's philosophy of return; as such, it is misleading. For it is the theory—called by Mach the indirect description—which is characterised by containing "nothing that is unessential" and which is abstract *par excellence*, and not the rich wealth of facts and much less abstract laws. Nevertheless, in these misleading terms, Mach is expressing a legitimate preference. He prefers comparative physics to atomic physics, and he is entitled to say so. In comparative physics there are, or can be, two orders of fact, in the study of both of which we have direct sense-perceptions. In the atomic physics, the one order does not permit

[5] W., p. 398. [6] P., p. 248.

direct sense-perceptions, for we have no direct sense-perceptions of the molecules and atoms. The atomic physics is strictly comparative also, but there is this definable difference.

If we accept for the moment Mach's account of science as the direct or indirect description of nature, we must also accept his requirement that the descriptions are to be made in the most economical terms possible:

> When the human mind, with its limited powers, attempts to mirror in itself the rich life of the world, of which it is itself only a small part, and which it can never hope to exhaust, it has every reason for proceeding economically.[7]

Any word exercises an economic function; but this economic function is magnified enormously when the word denotes one of the more profound metrical concepts of physics. The student who has mastered the idea of 'potential'[8] has the freedom of the entire territory of classical physics. There could hardly be a better example of the enormous power economically lodged in a single concept, or in a single word. If a single metrical concept represents an enormous economy, so much more does the law or experimental generalisation. *Prima facie* the simple algebraic formulae which express the laws of refraction of light and the falling body cover an indefinitely large number of cases, that is to say, a number larger than any finite number however large. Again, if a notable contribution to the economy of physics arises from a single metrical concept and from the comparatively simple law, it is not surprising that the economic value of a conceptual system, or of some extensive theoretical scheme, is very great indeed. It is required of such a scheme that it should contain the *smallest possible number* of independent judgements and that these judgements should be of the *simplest possible kind*; yet at the same time, from this minimum number of simplest possible judgements, it should be possible to deduce by logical inference all other judgements which are relevant to the given province of thought.

> The ideal of the economic and organic mutual adaptation of the compatible judgements belonging to a given province is reached,

[7] P., p. 186. [8] P., p. 197.

when one has succeeded in finding the smallest number of simplest independent judgements from which all the remaining judgements result as logical consequences, i.e. they can be deduced.[9]

In addition to economy *through* laws there is an economy *of* laws. I have already touched on the superiority of Galileo's account of the motion of a thrown ball as compared with Aristotle's account of the same.[10] The superiority lies in Galileo's economy *of* laws; he needs only one for the whole motion, whereas Aristotle needs two. The term 'law' is used loosely here for any kind of theory or mode of interpretation; laws are hypothetical, as theories are.

William of Ockham gave a general philosophical statement of the principle of economy, away back in the fourteenth century:

'Entia sunt non multiplicanda praeter necessitatem'
or 'Entities are not to be multiplied unnecessarily'.

If for 'entities' one writes in turn 'words', 'concepts', 'laws', 'theories', one has most of the various forms of Mach's idea of economy in Ockham's Razor.

In 1876, some years after the appearance of Mach's monograph on the *Conservation of Energy*, Kirchhoff published an authoritative textbook of mechanics. Kirchhoff like Mach sees the task of science as a matter of description, he notes that the description must be simple—this may be taken as a form of the economy idea—, and, again like Mach, he dislikes the notion of cause or concealed explanation:

. . . I maintain that the proper task of mechanics is to *describe* those motions which take place before us in nature, and of course to make this description as complete as possible and to make it in the simplest possible way. By that I mean that we are concerned here to declare *what* are the phenomena which take place, and not to ascertain their *causes*.[11]

It is not however easy to define 'simplicity':

There seems to be little agreement as to the importance of the so-called 'problem of simplicity'. . . .

[9] E., p. 179. [10] M., p. 171.

[11] G. Kirchhoff, *Vorlesungen über Mathematische Physik*, second edition, Leipzig: 1877, p. iii (first edition, 1876).

Until quite recently the idea of simplicity has been used uncritically, as though it were quite obvious what simplicity is, and why it should be valuable. Not a few philosophers of science have given the concept of simplicity a place of crucial importance in their theories, without even noticing the difficulties to which it gives rise. For example, the followers of Mach, Kirchhoff, and Avenarius have tried to replace the idea of a causal explanation by that of the 'simplest description'.[12]

When Mach, in the preface to the *Mechanics* (1883) refers to his "fundamental conception of the nature of science as Economy of Thought"[13] he goes much too far. Norman Campbell comments that the vast majority of folk attain economy of thought by the simple plan of not thinking.[14]

13. Difficulties

The descriptionist view of science, however true and important it may be in itself, cannot be the whole truth about science. This has already been noticed earlier in this chapter, and the views of Karl Pearson, Harold Jeffreys and Braithwaite have been cited. The main point is, that in no ordinary sense can a man describe what has not yet occurred, i.e. he cannot be said to describe the future. Men of science in fact are in possession of great theoretical systems which do much more for them than indirectly describe nature; they stimulate the creative genius and confer spectacular powers of prediction. For decades before matter was transformed into energy, the possibility of doing so was known, and the precise *mechanical equivalent of mass* was worked out. To Braithwaite's final question: "Is science invention or discovery?"[1]: Mach would have answered, 'Discovery'. All Mach's books defend this view of science in a brilliant and eloquent fashion. Yet Braithwaite's answer, 'Both', has more of the truth in it. It is often said that theories are free creations of the human mind, and this is at least a half-truth. The theory is the field for the exercise of man's inventive genius. It is discovery and experiment which falsify or verify the theory.[2]

[12] K. R. Popper, *The Logic of Scientific Discovery*, p. 136.
[13] M., p. xxiii. [14] N. R. Campbell, *Foundations of Science*, p. 222.
[1] R. B. Braithwaite, *Scientific Explanation*, p. 367. [2] Ibid., p. 368.

There are other serious objections to Mach's views. For example, *experience* does not become *science* until it is *communicated*. But what is communicated cannot be *my* experience. For the communication to be valid science, it must be unambiguous, there must be no mistake about it, and it must become publicly available. This requirement stands although no one can prove that *my* experience of red is not *your* experience of yellow. P. Alexander has discussed the consequences of these obvious facts in an important book:

It is, of course, essential for obtaining the communicability necessary for science that we find basic *statements*; experiences will not do. Sensationalists must therefore hold that it is proper to talk about both elementary *experiences* and elementary *statements* as things about which we cannot be mistaken.[3]

Properly understood, this is in itself sufficient to refute the theory that science is or can be merely a description of sense-perceptions:

Suppose that I . . . say 'I am now sensing red.' This could be an incorrigible statement but only if it is purely autobiographical. If it is to be more than this then just because it makes a public claim it would be corrigible since it is open to others to say 'It's not red, it's green.' As soon as it ceases to be merely autobiographical it appeals to a public standard and can be disputed by anyone on the grounds of failure to come up to that standard.

According to his principles, the sensationalist must always agree, or can never disagree, if I say 'I am sensing red'. But this agreement is odd. It is not agreement upon the similarity of our sensations but an admission that if I say it he has no grounds for disputing it, and *can* have no grounds. This alone is enough to render the sensationalist account of science untenable.[4]

No one can dispute *my* sense-perception, and it is incorrigible in this sense. But it is insulated from everyone else, and therefore the "requirement of publicity is incompatible with the requirement of incorrigibility."[5] It is somewhat beyond the scope of this book to take this much further. But the neo-positivists of the present time have tried to meet the objection:

[3] P. Alexander, *Sensationalism and Scientific Explanation*, London: 1963, p. 41. [4] Ibid., p. 43. [5] Ibid., p. 44.

> The truth of statements can not be guaranteed through experience, for statements to be scientific must be intersubjective, and their value can only be established on an intersubjective basis, and not through experience which is subjective.[6]

In popular lectures on the philosophy of science, I have often discussed the physicist's preference for a numerical wave-length to a colour—red for example—and I have suggested that in this way the enduring heritage of Cartesianism is manifested. It could more simply be that a statement of wave-length is 'intersubjective' or 'public' whereas a colour can only be a private experience. The gist of Kraft's book amounts to the thesis that Mach's 'elements' have to be replaced by intersubjective statements (die intersubjectiven Aussagen). Mach was therefore misguided when he advised physicists to return from the indirect to the direct description of nature; evidently, in general, scientific theory is intersubjective and is therefore more properly the goal of science than the experiences of individuals.

Alexander points out the ambiguity of the term 'description'.[7] Whatever is meant by the 'scientific description' of a celandine, it cannot be a heightened form of the descriptions employed in everyday life. Suppose we were to give spatial (x, y, z) and temporal (t) coordinates of each patch of yellow (taken as an 'element') and each patch of green (another 'element'), in the end we should achieve a description of enormous complexity *which would have very little interest to the scientist, as such, at all.* Indeed, such a description would have no interest to anyone. This is sufficient to refute Mach's statement that it is necessary to substitute indirect by direct description; evidently the scientist as such does not *retreat from* the structural formula of chelidonic acid *back to* the space and time coordinates for yellow patches. In the ordinary practice of science the substitution of indirect by direct description is neither advisable nor necessary. The term 'indirect' is a logical escape-hatch for Mach's interpretation of science. If the meaning of 'indirect' is sufficiently extended, then the statement that 'science is the indirect

[6] V. Kraft, *Der Wiener Kreis,* Wien: 1968, p. 114.

[7] P. Alexander, *Sensationalism and Scientific Explanation,* p. 112.

description of natural or contrived phenomena' could be true. In the terms of this essay:

i. Any description satisfactory to a scientist involves 'non-O' concepts;

ii. Therefore no satisfactory description can be direct.

Even so Mach's extremely valuable analysis of the 'non-O' concepts, in particular his 'reduction' of electrical charge to mechanical force ('O' concept) and his definition of mass ('non-O') in terms of accelerations ('O'), stands untouched by Alexander's criticism.

Alexander's book is by no means entirely directed against Mach's views. He makes the good point, that if there are rival theories for the interpretation of certain areas of experience, these areas being in some sense analysable into sense-perceptions and pointer-readings, "there must also be non-rival descriptions or else the theories could not be rivals."[8] Alexander's book has a red cover and the description 'red' in this sense is absolute. For if that which is being interpreted by theory changes during the process of interpretation, there can be no interpretation.

> If scientists set out to explain how things have that property to which we use 'red' to refer, then they fail in their task if one result of their work is to change the signification of the term; what they explain is something different from what they set out to explain.[9]

He adds a further telling example. 'Temperature' means in the first place exactly the same to the scientist as it does to the layman; it is either 'Wärmezustand' or even 'Wärmeempfindung'. *And it must go on meaning this,* if the scientific quest is to be intelligible. Mr. Alexander's protest that temperature "does not *mean* . . . for the scientist 'mean kinetic energy of molecular motion' or, indeed, anything different from what it means for the layman; . . ."[10] is worthy of Mach himself.

14. The greatness of Mach

Looked at from our present point of vantage, physical science is a system of theories (Popper). But Mach resolutely declines to regard science in this way. In his company, we must look

[8] Ibid., p. 92. [9] Ibid., p. 95. [10] Ibid., p. 96.

behind the theory to the concept, behind the concept to the experiment, and behind the experiment to the common experiences of children and primitive tribes; also with Mach we must admit that the rich sensuous experience of mankind can be broken down into sense-perceptions.

What then is the main problem in the philosophy of science? It is, according to Mach, the transmutation of experience into theory. And Mach illuminates this question, at least partially. The units or elements of scientic theory are the metrical concepts. Some of these—temperature and force are the best examples—are in a special direct relationship to sense-perception. In the simple terms of Hume, force and temperature can almost be *felt*. There is here therefore a genuine link between the realms of experience and systematisation.

It is true that many other metrical concepts of physical science lack this direct relationship back to sense-perception, and that these 'indirect' (mittelbar) concepts are amongst the most powerful counters in the advanced theories of physical science. I refer to mass, energy, entropy, wave-function and many others. Mach however reminds us that temperature and force (unmittelbar) are defined, in classical physics, through energy and mass (mittelbar) respectively.

It is not true that the man of science is *entirely* free as he builds up his theory. He is, and ought to be, constrained and checked all the while by the whole of past and present experience. Science may be like art, but it is even more like history. Mach again reminds us that an air-borne phantasy with the life-line back to the laboratory bench severed, is of no value to science.

So Mach's three textbooks are historical in tone. What *has been* valuable in man's experience of nature, *is* valuable still. The aim is always to *know* nature.

This is so fundamental that Mach rarely moves on to secondary questions—the nature of explanation, the problem of induction and so on. For after all, science is just the *indirect description of nature*.

Ernst Mach: a biographical note[1]

Ernst Mach was born on 18 February 1838, in Turas, Moravia. His father, Johann Mach, was a teacher, and later a smallholder. His mother, unworldly like her husband, was interested in music and art. In the main, Ernst was educated at home by his father. At the age of 15 he entered the grammar school at Kremsier, Moravia.

Two years later, he proceeded to the University of Vienna where he read mathematics and physics. As Privat-Dozent, he taught physics from 1861 to 1864.

In 1864, Mach was called to the chair of mathematics in the University of Graz, Austria. The chair of physics was added later. Important papers on the psychology of sense-perceptions began to appear about this time. Mach's very extensive investigations on this subject are held in high esteem by psychologists, and they constitute the point of departure in his philosophy of physics.

Mach became professor of physics at the German university of Prague in 1867, which was also the year of his marriage to Ludovica Marrusig. Also in 1867, appeared his important paper *On the Definition of Mass* in Carls Repertorium. This was reproduced and amplified in the monograph on *The Conservation of Energy* (1872). Mach's textbook of mechanics, treated as a branch of physics, was published in 1883. It is his best and most important work.

His experimental work on the photography of sound waves began in 1884. Mach is known to many through the modern

[1] H. Henning: *Ernst Mach als Philosoph, Physiker und Psycholog*, Leipzig: 1915, pp. 1–8, pp. xi–xviii. H. Dingler, *Die Grundgedanken der Machschen Philosophie*, Leipzig: 1924, pp. 8–16. F. Ratliff, *Mach Bands*, San Francisco: 1965, pp. 8–20. K. D. Heller, *Ernst Mach*, Wien: 1964, pp. 1–22, pp. 96–8, pp. 133–43.

practice of calling the ratio of airspeed to the speed of sound by the term 'Mach number'. When a projectile or aircraft travels at such a speed that the Mach number exceeds unity the bow-wave is formed. Others know Mach as the physicist whose writings influenced Einstein.

In 1895 Mach returned to the university of Vienna as professor of the 'History and Theory of Inductive Science'. 1896 saw the publication of the *Principles of Heat* and also a collection of *Popular Lectures*. *Knowledge and Error* was published in 1905. It exhibits the consequences of combining Hume's ideas on the methodology of science with an expert knowledge of physics, psychology and mathematics.

Mach suffered a stroke (1898) which intermittently crippled the right side of his body. He retired in 1901, and continued to live near Vienna until 1913. During this time he prepared his last textbook, the *Principles of Physical Optics*. Mach did not live to see the book published, nor to write the projected second volume on the same subject.

His last three years were spent at the home of his son, Ludwig Mach, in Munich-Vaterstetten. Einstein visited him in 1913. Ernst Mach died on 19 February 1916, the day after his 78th birthday.

Index of names

More important references are italicised

Index of subjects

More important references are italicised